Fossils in Focus

About the Authors and Respondent

J. KERBY ANDERSON

Kerby Anderson is a lecture and research associate with Probe Ministries International. He obtained his B.S. in zoology from Oregon State University and his M.F.S. in ecology and evolution from Yale University. He is a member of Blue Key National Honorary and was awarded the U. G. Dubach award for outstanding leadership while at Oregon State University.

HAROLD G. COFFIN

Harold G. Coffin is professor of paleontology at the Geoscience Research Institute and at Andrews University, Berrien Springs, Michigan. He obtained his B.A. and M.A. degrees in biology at Walla Walla College and his Ph.D. in zoology at the University of Southern California. Dr. Coffin is a member of the American Association for the Advancement of Science, the Geological Society of America, and Sigma Xi. He has had papers published in the *Journal of Paleontology* and the *Bulletin of the Geological Society of America*. He is also the author of two books *Creation: Accident or Design* and *Earth Story*.

RUSSELL L. MIXTER

Russell L. Mixter is professor of zoology at Wheaton College, Wheaton, Illinois. He holds a B.S. degree from Wheaton, an M.S. degree from Michigan State University, and a Ph.D. in anatomy from the University of Illinois.

A frequent participant in the National Science Foundation's summer programs, Dr. Mixter has directed an NSF Institute for elementary teachers. He is past president of the American Scientific Affiliation and former editor of the ASA Journal. He edited the book *Evolution and Christian Thought Today* and has authored a monograph *Creation and Evolution*, both of which were published by the ASA.

Fossils in Focus

J. Kerby Anderson
and
Harold G. Coffin

with a response by
Russell L. Mixter

Third printing 1981

Library of Congress Cataloging in Publication Data

Coffin, Harold G
Fossils in focus.

(Christian free university curriculum)
Bibliography: p.
1. Paleontology. 2. Geology—Philosophy.
3. Bible and evolution. I. Anderson, Kerby, joint author. II. Title.
QE721.C72 560 77-9536

ISBN 0-310-35741-1

Place of Printing *Printed in the United States of America*

Design Cover design by Paul Lewis
Book design by Louise Bauer

Permissions

pages 34,35	From *The Science of Biology* by Paul Weisz. Copyright 1971 by McGraw-Hill Book Company. Used with permission of McGraw-Hill Book Company (figures 2,3).
pages 42,43	From *Paleobiology of the Invertebrates* by Paul Tasch. Copyright 1973. Courtesy of John Wiley and Sons, Inc. (figures 5,6).
pages 51,54, 63,67	From *Vertebrate Paleontology* by Alfred S. Romer. Copyright 1966. Courtesy of the University of Chicago Press (figures 7,8,10,13, Basilosaurus and Kentriodon).
page 56	From *Evolution of the Vertebrates* by Edwin H. Colbert. Copyright 1969. Courtesy of John Wiley and Sons, Inc. (figure 9).
page 64	From *Biology of Bats* edited by William Wimsatt. Copyright 1970. Courtesy of Academic Press (figure 11).
page 65	From *The Life of Vertebrates* by J. Z. Young. Published by Oxford University Press (figure 12).
page 67	From *Evolution Emerging* by William King Gregory. Copyright 1951. Courtesy of the American Museum of Natural History (figure 13, Balaenoptera physalus).
page 71	From *Morphology and Evolution of Fossil Plants* by Theodore Delevoryas. Copyright 1962. Holt, Rinehart and Winston. Used by permission of Senckenbergische Naturforschende Gesellschaft (Frankfurt, West Germany). Originally appeared as "Beitrage Zur Kenntnis der Devonflora II" by R. Krausel and H. Weyland in *Senckenbergana Biologica*, 40,113-115 (figure 14).

What is the Christian Free University Curriculum?

The Christian Free University Curriculum is a continuing series on course topics and current issues relevant to today's academic community. The authors and publishers of the Curriculum, while upholding the necessity of a clear separation of church and state *as institutions*, have designed the Curriculum to contribute to the academic process and to the preservation of true academic freedom in the realm of educational ideas.

The series has been developed in a scholarly, nonsectarian way for use as classroom collateral reading. Inherent in its preparation is a recognition of the goal of higher education to consider a breadth of viewpoints in the quest for truth, understanding, and values.

The Christian Free University Curriculum is published jointly by Zondervan Publishing House and Probe Ministeries International. Probe Ministries is a nonprofit corporation organized to provide perspective on the relationship of the academic disciplines and historic Christianity. Probe members and associates are actively engaged in research and in lecturing with students and faculty in thousands of university classrooms throughout the United States and Canada.

Christian Free University books should be ordered from Zondervan Publishing House (in the United Kingdom from The Paternoster Press), but futher information about Probe's materials and ministries may be obtained by writing to Probe Ministries International, Box 5012, Richardson, Texas 75080.

Contents

Illustrations

Book Abstract

The fossil record has traditionally been presented as the foremost historical evidence for the theory of evolution. The Neo-Darwinian model, however, does not fit the facts of paleontology. In this study, various models of origins are considered (Neo-Darwinian, Saltation, Punctuated Equilibria, Special Creation) and the strengths and weaknesses of these models are evaluated in light of the fossil record of invertebrates, vertebrates, and plants. The creation model is shown to be the model that fits the data of paleontology most directly.

Chapter Abstract

If there is any historical evidence for evolution, it can best be found in the fossil record.

chapter one

The Question of Origins

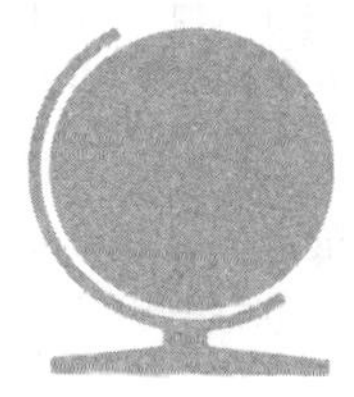

Man is unique in the animal kingdom. He is aware of his own existence and he is curious about the past, present, and future. Though animals show some curiosity about their present situation, only man is curious about both his origin and his future.

Through the ages, societies have proposed a number of ways to explain beginnings. In recent times, the theory of evolution has been proposed as the explanation for the origin and development of life. It has been perhaps the most influential scientific theory of the nineteenth and twentieth centuries. Not only has it had a marked effect on the biological and physical sciences but it has greatly influenced other academic disciplines as well. At the heart of each academic discipline is a basic consideration of the question of origins. To ask about man and the world around him is to ask the basic question about life on earth. The view that man has concerning himself and the world around him has been greatly influenced by the theory of evolution.

The reality of evolution is taken for granted by most

of the scientific community, and investigations are directed toward understanding the processes and mechanisms of evolution. In recent years, however, a growing number of individuals have begun to question the validity of the theory of evolution or some of its postulates. Their challenge has been raised on philosophical and scientific grounds.

The Great Debate

There are qualified scientists on both sides of this question. Each group can produce evidence for or against the theory of evolution. This evidence has come from areas such as comparative anatomy,[1] biochemistry,[2] population genetics,[3] and biogeography.[4]

As a part of this debate about the theory of evolution, there has been some controversy surrounding the attempts to explain the origin of life from inorganic compounds. There is a growing body of scientists claiming that these proposals, because of their mathematical and biophysical improbability, are inadequate to explain the origin of life. This body of scientists includes both evolutionists[5] and nonevolutionists.[6]

All of this rhetoric is certainly interesting and often entertaining, but it fails to answer our basic question. How did life on this earth arise? Research can only provide circumstantial evidence — no direct evidence can be obtained to show that evolution *did* indeed occur.

There is general agreement among most scientists that the only possible source of *historical* evidence for evolution is the fossil record. In a real sense, fossils constitute the last court of appeal concerning the history of life.

> Although the comparative study of living animals and plants may give very convincing circumstantial evidence, fossils provide the only historical documentary evidence that life has evolved from simpler to more complex forms.[7]

Thus, if we are to answer the question at hand satisfactorily, it would be good to confine our investigation to an examination of the fossil record.

Although fossils provide a record of ancient life, this does not mean that all forms of past life can be found. The fossil record is a relatively incomplete representation of the past. No doubt many ancient forms of plant and animal life have yet to be discovered.

Sharpening the Focus on the Fossil Record

This situation poses a number of problems. Scientists must in some way interpret the available data in order to make sense of the information. This process of interpretation is subjective rather than objective and can lead to a number of problems.

> The actual data, then, normally consist of relatively small samples of the lineage, scattered more or less at random in space and time. The process of interpretation consists of connecting these samples in a way necessarily more or less subjective, and students may use the same data to "prove" diametrically opposed theories.[8]

The process of interpreting the fossil record is made more difficult if the observer is subject to any bias. The analysis of each piece of information can be hindered when only the evolutionary hypothesis is considered in interpreting the data.

> Thus the paleontologist can provide knowledge that cannot be provided by biological principles alone. But he cannot provide us with evolution. We can leave the fossil record free of *a theory of evolution*. An evolutionist, however, cannot leave the fossil record free of the *evolutionary hypothesis*.[9]

It is important that a scientist avoid circular reasoning in his investigations. But this is difficult today when the only theory of origins being considered by the scientific community is the theory of evolution.

In this book we shall consider a number of models for origins that have been advanced and compare them with the evidence from the fossil record. An examination of the various models simultaneously should provide a more objective standard for judging the fossil record. The various theories presented should stand or fall on this evidence.

Chapter Abstract

Various models of origins are presented to provide a framework for an investigation of the fossil record of plants and animals. Tests are proposed that will help us to determine the validity of each model.

chapter two

Models for Origins

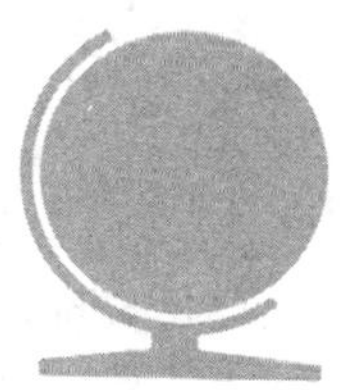

In the beginning of this chapter it should be mentioned that none of the models for origins strictly qualifies as a scientific theory. Each explanation postulates that some mechanism in the past produced life. Since this mechanism took place in the past, it cannot be taken into a laboratory today and subjected to rigorous testing and observation. It cannot be repeated; thus, it is outside of the standard methods used in scientific experimentation. As a result, we cannot ultimately prove or disprove any of these explanations with the same rigor that we normally use in the method of science. The approach used here will be to describe the principal models of origins and consider them in light of the fossil record.

Neo-Darwinian Evolution

The classic model for evolution is the one originally proposed by Charles Darwin. The modern form of this original model is referred to as the Neo-Darwinian theory of evolution.[10] In this modern form, the basic mechanisms for the origin of new species are the processes of mutation and natural selection.

This model postulates that evolution took place at the species level. Between each species found today, we should expect to find numerous transitional forms in the past that would link these different species to a common ancestor. Evolutionists such as George Gaylord Simpson find it "nearly impossible to imagine these [evolutionary] processes occurring except by transition over a long sequence of generations."[11]

The search for these transitional forms (missing links) by paleontologists has not been very successful. Each major group of organisms appears abruptly in the fossil record without any transitions.

> It is a feature of the known fossil record that most taxa appear abruptly. They are not, as a rule, led up to by a sequence of almost imperceptibly changing forerunners such as Darwin believed should be usual in evolution.[12]

> Despite the bright promise that paleontology provides a means of "seeing" evolution, it has presented some nasty difficulties for evolutionists, the most notorious of which is the presence of "gaps" in the fossil record. Evolution requires intermediate forms between species and paleontology does not provide them.[13]

The lack of intermediate forms is no small problem for this model. The Neo-Darwinian theory of evolution maintains that such missing links should be found. Over a hundred years ago, Darwin suggested that these gaps existed because of an imperfect fossil record.

> Why then is not every geological formation and every stratum full of such intermediate links? Geology assuredly does not reveal any such finely graduated organic chain; and this, perhaps is the most obvious and serious objection which can be urged against the theory. The explanation lies, as I believe, in the extreme imperfection of the geological record.[14]

A century of work in paleontology has failed to produce the fossils necessary to confirm this model. In fact, Norman Newell of the American Museum of Natural History states, "Experience shows that the gaps which separate the highest categories may never be bridged in the fossil record. Many of the discontinuities tend to be more and more emphasized with increased collecting."[15]

Saltation and the Hopeful Monster

The failure of the Neo-Darwinian model to adequately explain the persistent absence of transitional forms has stimulated other evolutionists to propose alternative theories of evolution. One serious attempt was made by Richard Goldschmidt[16] and Otto Schindewolf.[17] They suggested that evolution took place through massive evolutionary changes. These sudden jumps (saltation) were a result of a series of mutations.

Their model postulated two types of mutation. Geneticists observed small mutations called micromutations that accumulate over generations to generate new races or species. Goldschmidt and Schindewolf proposed that there are also large mutations called megamutations that generate larger orders of classification in a few generations.[18] These sudden jumps in evolution would produce an entirely new creature, the so-called "hopeful monster." In other words, this model envisions a reptile that would lay an egg that would hatch into a bird.

This theory is disregarded by most evolutionists today.[19] Most feel that it is Goldschmidt and Schindewolf who laid the egg. Furthermore, most scientists do not believe that such a mechanism, even if it did exist, could account for the complex anatomical and physiological adaptations that would have to evolve simultaneously in order to achieve such functions as running, swimming, or flying.

Punctuated Equilibria

The newest modification of the traditional evolutionary theory has come from paleontologists who have applied the principles of biology to the data from paleontology. Present biological theories of evolution predict that species are formed through a process known as allopatric speciation.[20] According to this model, a small number of species A are isolated from the parent population by a natural barrier, thus estab-

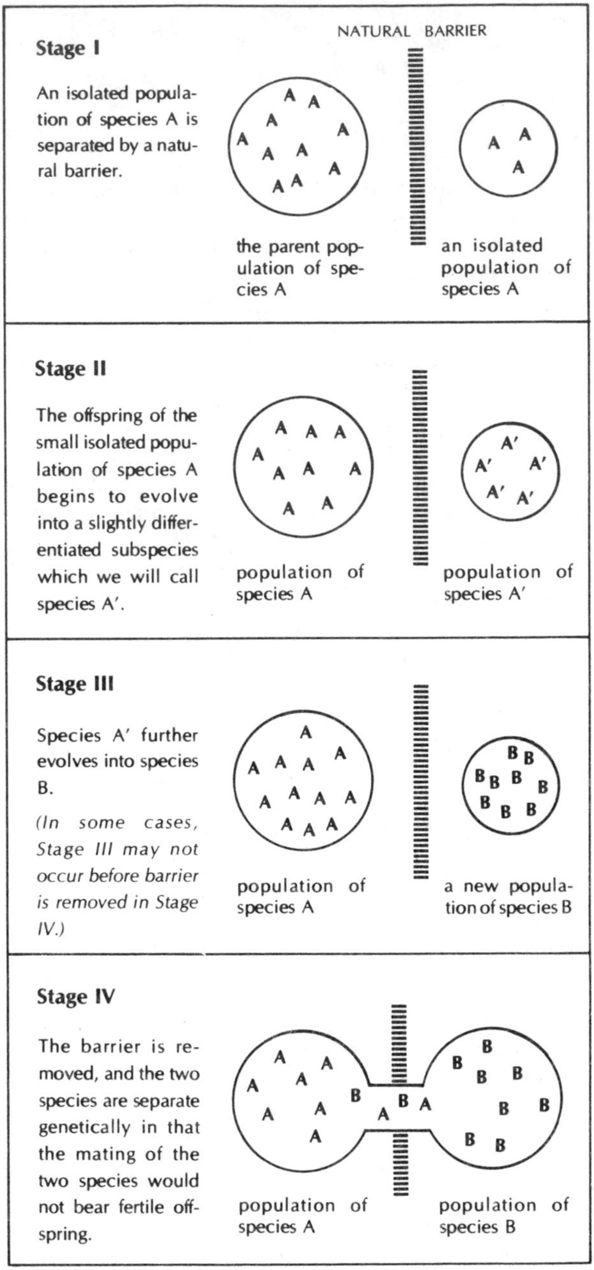

NATURAL BARRIER

Stage I

An isolated population of species A is separated by a natural barrier.

the parent population of species A

an isolated population of species A

Stage II

The offspring of the small isolated population of species A begins to evolve into a slightly differentiated subspecies which we will call species A′.

population of species A

population of species A′

Stage III

Species A′ further evolves into species B.

(In some cases, Stage III may not occur before barrier is removed in Stage IV.)

population of species A

a new population of species B

Stage IV

The barrier is removed, and the two species are separate genetically in that the mating of the two species would not bear fertile offspring.

population of species A

population of species B

Figure 1. The postulated mode of allopatric speciation.

lishing two populations. Then the small isolated population evolves into a new species (see figure 1). This new species would evolve through the processes of mutation and natural selection. Since these two groups are isolated, different mutations and different pressures of natural selection would operate on the two groups to form a new species that is genetically different from the parent population.

Paleontologists such as Niles Eldredge and Stephen Jay Gould[21] have adopted this model from biology and have applied it to the field of paleontology. They have noted, as have other evolutionists, that gaps in the fossil record seem to be the consistent rule. Thus, they have formed this model from a synthesis of biology and paleontology. They argue that their model would predict the regular and systematic gaps we find in the fossil record. Small, isolated populations would evolve intermediate steps faster than the parent population and would therefore be less represented in the fossil record.

There are a number of flaws in this model that should be considered here. First, this model postulates a mechanism that would probably account for only minor changes among species. The isolation of populations and the development of variations due to mutations might explain why we would not find a smooth continuum of steps leading from one fossil species to another. However, it would not explain why we do not find occasional transitional forms bridging the gap between major groups of plants and animals.

Second, it requires that we believe these isolated populations left an insignificant number of individuals behind in the fossil record. But, in the evolution of higher orders of classification, we would expect to find some traces of transitional species. Between each of these higher orders of classification, we would predict that there would be a number of transitions. And each of these transitional species would encompass hundreds of generations of plants or animals. And again, within each of these generations of a transitional species we would expect to find numerous individuals. The fact that we do not find any transitional species that bridge the gap between these groups certainly is a very curious phenomenon.

Finally, it is apparent that this model is not in agreement with the studies of population genetics. Scientists who have studied the genetics of populations have not been able to see major evolutionary changes taking place in populations that would correspond to the types of changes necessary for evolution to occur. It has been traditionally stated that evolution cannot be observed in modern times, since it happens slowly over a period of millions of years. However, now we have paleontologists telling us that the reason we do not see evolution in the fossil record is because of the rapid evolution of isolated populations during these same millions of years. In other words, one group is telling us that the reason we do not see evolution is that it happens too fast. And the other group is telling us that the reason we do not see evolution taking place is that it happens too slowly. Are these two positions incompatible?

The Creation Model

In recent years, a growing body of scientists has suggested that the most reasonable explanation for the fossil record cannot be found in the models of evolution. Instead, they suggest that the fossil record can best be explained by some model of creation.[22] According to this model, all major "kinds" of organisms were created individually. These created kinds would have the genetic variability to adapt to different habitats and environments.

The modern creationist does not deny the existence of change in various organisms today. Certainly in modern times there is an abundance of evidence of limited change. Man has been able to produce changes within a great number of plant and animal types. The creationist, however, is unwilling to extrapolate the evidences for minor change to major change. In other words, the creation model would predict that there would be a fixity of kinds and we would not expect to find transitional forms from one basic kind to another (annelids to arthropods, fish to amphibians) as the result of the accumulation of many small changes.

Experimentation in genetic laboratories has not succeeded in producing any major changes in plants or animals. For example, many types of fruit flies have been developed in the laboratory but all of these types are only different varieties or species within the same genus. They have not changed from one basic type of fly into another basic category. There appears to be limits to the amount of change possible within kinds.

Attempts to reduce the number of bristles on the body of the fruit fly ended when the number dropped to twenty-five from the normal average of thirty-six, because these flies became sterile. Attempts to increase the bristle number were only a little more successful, but again the experiment was stopped for the same reason when there were fifty-six bristles.[23] Despite the efforts of science for the past one hundred years, plants and animals have not been pushed out of their basic categories. It is difficult to believe that the chance processes of nature would be able to accomplish what intelligent man has not been able to do in his multimillion-dollar laboratories.

It is interesting to note that a misunderstanding concerning change actually led to the formation of Charles Darwin's theory of evolution. He had been taught that species were fixed, that there was no change. During the voyage of the *Beagle,* Charles Darwin observed powerful evidence that species did indeed change. The varieties of birds and other animals on the Galapagos Islands caused him to change his thinking about the fixity of species. However, he did not observe any evidence during his voyage or on the Galapagos Islands that forced him to the conclusion that change was limitless. Change is possible within limits. In fact, as we look at the fossil record, the results from genetic research, and the natural world about us, we are led to believe that the truth lies between the two extremes of fixity of species and limitless change.

Testing the Models

Before we begin our examination of the fossil record, it would be appropriate to make a few comments about the kinds of tests we would like to devise in order to test the validity of these various models of origins.

Since each of the models of origins would allow for minor variations of organisms at lower levels of classification, it would be of little value to consider these lower levels. In testing the validity of the models, we should, instead, concentrate our examination on the fossil relationships that have been proposed between higher levels of classification.

We can imagine that we have before us a collection of various fossil organisms that we would like to place in a predictable pattern. As we begin to organize this collection, we can imagine that there would be certain questions we would like to ask in order to make our job less difficult. One of the major limitations of the fossil record is that we cannot ask certain questions that would be helpful in our task.

One of the first questions we would like to ask is whether similarities between organisms are a result of a direct relationship (kinship) or a result of common design. Quickly we would note that whales and fish, for example, were very similar. It would be helpful for us to know whether this similarity is due to a close relationship (i.e., one may have evolved from the other) or due to common design for a common function (i.e., both need to survive in water). In this case, both the creation and evolution models would agree that the similarities are a result of common design due either to convergence (evolution models) or creation (creation model).

In other areas of the fossil record, our identification would be more difficult. In some cases, common structure (comparative anatomy) or common development (ontogeny) might indeed imply a common ancestry and a direct relationship (phylogeny). However, in other cases, a common structure or development would not imply a direct relationship. As we identify potential transitional forms, it is important that we be careful not to make similarity in structure our sole proof for a common evolutionary relationship.

We should also be aware that although two organisms may share the same genetic constitution (genotype), they may exhibit a marked difference in anatomy (phenotype). In fact, John Imbrie of Columbia University has said that a "troublesome limitation of paleontological data lies in the difficulty (usually the impossibility) of distinguishing phenotypic from genotypic variation."[24] For example, we find many varieties of dogs within the dog "kind." Although these varieties are all essentially identical genetically (genotypically), they appear visually (phenotypically) to be very different. Imagine if we were to find a dachshund, a beagle, a collie, and a great dane in the fossil record. Although we could construct a phylogenetic relationship (evolutionary tree) leading from a dachshund to a great dane, it would not be a true representation. The great dane did not evolve from a dachshund. Both of these dogs are simply variations of a single dog "kind" with the same genetic structure.

As we consider transitional forms between higher levels of classification, we should note the similarities, the differences, and the position in the geologic column of each fossil find. As we begin to determine its relationship to other fossils, we should be aware that all of these characteristics are important in making a sound judgment on any possible evolutionary relationship. If we ever find a good sequence of transitional fossils, this would certainly be evidence for one of our models of evolution. If we find a consistent presence of unbridged gaps (discontinuities) between these groups, that would be evidence for the creation model.

Chapter Abstract

The fossil record of invertebrates is considered in light of the various models of origins. Particular reference is made to the appearance of most of the major invertebrate phyla in the Cambrian period.

chapter three

The Invertebrate Fossil Record

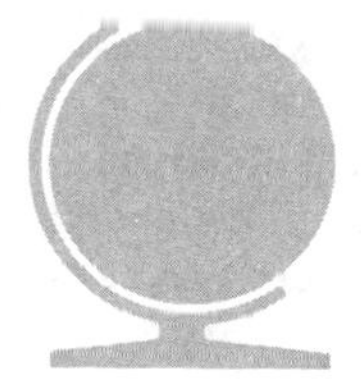

According to the evolutionary model, one-celled life originated at some time in the past from inorganic substances. Simple, one-celled forms of life evolved into more complex forms through time. The vast majority of all these forms were invertebrates (animals without backbones). Thus it is most reasonable to begin our investigation with the fossil record of the invertebrates.

An investigation of invertebrate fossils suffers from some problems. One major problem is that many of them do not leave conspicuous and identifiable fossils. Invertebrates are often smaller than other plant or animal groups. Small fossils are more difficult to find and they have a greater potential for being completely destroyed by geological forces. This difficulty may find compensation in the greater abundance of smaller specimens.

Some invertebrates are primarily composed of soft parts and seldom leave a fossil record. Those invertebrates that have preservable hard parts (such as a shell) are found much more frequently. Therefore, we would expect that certain invertebrate groups will be better represented by fossils than others. Our validation of origin models should be done with portions of the fossil record that are paleontologically well represented. Fossils of such groups can be expected to document their origins and subsequent history.

The Cambrian Explosion of Life

At the time Charles Darwin wrote *The Origin of Species,* there were already a number of difficulties with his theory. One major difficulty was this presence of gaps in the fossil record. A second major difficulty was the abrupt appearance of complex animals in the Cambrian rocks, without evidence of simpler ancestors in older rocks.

The Cambrian period is said to be the oldest period in which a significant number of complex fossils can be seen. Rocks below the Cambrian are usually devoid of evidences of living organisms and may be lumped together under the heading of Precambrian. Although the Precambrian rocks contain very few fossils, the Cambrian sediments show a large array. Almost all of the major phyla of invertebrates are represented in Cambrian rocks. The lower Cambrian contains these phyla:

Protozoa—single-celled animals
Porifera—sponges
Coelenterata—jellyfish, anemones, and corals
Archeocyatha—sponge or corallike creatures now extinct
Brachiopoda—bivalves superficially resembling clams (lampshells)
Mollusca—clams, snails, octopuses
Arthropoda—jointed-leg creatures, such as insects, crustaceans, and trilobites
Echinodermata—spiny-skinned animals, such as starfish, sea urchins, sea lilies

At least two additional invertebrate phyla (Chaetognatha — arrowworms — and Annelida — segmented worms) though less abundant, also could be included if all of the Cambrian were considered.[25] In fact even the vertebrates (phylum Chordata) have been reported for the Cambrian, although most paleontologists do not allow for chordates so low in the geological column.[26]

Over one hundred years ago Charles Darwin could provide no satisfactory answer to this problem:

> To the question why we do not find rich fossiliferous deposits belonging to these assumed earliest periods prior to the Cambrian system, I can give no satisfactory answer . . . The case at present must remain inexplicable; and may be truly urged as a valid argument against the views here entertained.[27]

If the theory of evolution is correct, the Cambrian fauna should be preceded by a series of increasingly complex ancestors. In Charles Darwin's day, paleontology was a young science. It was his belief that continued search for fossils would reveal the ancestors to the amazing array of intricate Cambrian animals. But modern paleontologists continue to find the problem that Darwin pinpointed. In 1959, at the centennial celebration of the publication of Darwin's famous book, Normal Newell of the American Museum of Natural History summarized the results of Precambrian research to that time:

> A century of intensive search for fossils in the pre-Cambrian rocks has thrown very little light on this problem. Early theories that those rocks were dominantly nonmarine or that once-contained fossils have been destroyed by heat and pressure have been abandoned because the pre-Cambrian rocks of many districts physically are very similar to younger rocks in all respects except that they rarely contain any records whatsoever of past life.[28]

It is generally acknowledged by most paleontologists that there is no evidence of ancestral forms for Cambrian fossils.

> Cambrian fossil animals, all marine invertebrates, certainly are not the remains of primitive organisms. [The trilobites] are highly organized complex arthropods and represent one of the most advanced great animal groups or phyla. Other fossils indicate that all the principal animal phyla inhabiting the world today probably were in exist-

> ence before the beginning of the Cambrian Period. None of these creatures stood anywhere near the base of the evolutionary ladder, and they provide no evidence concerning the nature of the first living things.[29]

Because such evidence is crucial for the support of the theory of evolution, a great deal of exploration has been undertaken in Precambrian sedimentary rocks since Darwin's time. The problem is more severe now because continued collecting has only emphasized that complex Cambrian animals world-wide were *not* preceded by simple Precambrian ancestral forms.

Since the publication of *The Origin of Species,* many claims of Precambrian fossils have been made. Some of these remain valid, but many have been discarded as misidentifications or pseudofossils. In recent years, intensive research on Precambrian sediments has been undertaken. Scientific papers have appeared, describing microscopic plant forms such as bacteria, fungi, and algae.[30] Other so-called Precambrian assemblages such as the Ediacara fauna of Australia are close to the Precambrian-Cambrian contact and either are very late Precambrian or early Cambrian forms.[31] Preston Cloud of the University of California at Santa Barbara insists that "in spite of many published records, there are as yet no unequivocal Metazoa [multicellular animals] in rocks of indisputable Proterozoic [Precambrian] or older age."[32] The important question is not, Are these valid fossils? but rather, Are these genuine ancestors or precursors to the Cambrian fauna? Are they indeed true evolutionary transitions between the complex Cambrian forms and the one-celled forms evolutionists assume existed in the Precambrian era? Most paleontologists will readily admit that the evidence to date from Precambrian rocks is poor and makes no significant contribution to explaining the origin of Cambrian animals.

Various attempts have been made to answer this question. Daniel Axelrod, in a paper on "Early Cambrian Marine Fauna," has listed a number of possible explanations. Those most often seen in the literature are these:

1. Precambrian animals had no preservable skele-

tons because there was little or no calcium in the sea water.

2. Precambrian organisms were seashore animals — the seashore being an active environment not likely to preserve the animals.
3. Organisms changed by major jumps of evolution at the beginning of the Cambrian era.
4. Precambrian organisms were too small to form readily seen fossils.
5. The Cambrian layer represents much time; consequently the change is not as sudden as often thought.

The same paper by Axelrod has in its introduction this informative statement concerning the Precambrian problem:

> One of the major unsolved problems of geology and evolution is the occurrence of diversified, multicellular marine invertebrates in Lower Cambrian rocks on all the continents and their absence in rocks of greater age. These Early Cambrian fossils include porifera, coelenterates, brachiopods, mollusca, echinoids, and arthropods. In the Arthropoda are included the well-known trilobites, which were complexly organized with well-differentiated head and tail, numerous thoracic parts, jointed legs, and — like the later crustaceans — a complex respiratory system. . . . Their high degree of organization clearly indicates that a long period of evolution preceded their appearance in the record. However, when we turn to examine the Precambrian rocks for the forerunners of these Early Cambrian fossils, they are nowhere to be found.[33]

None of these suggestions have been well accepted by a majority of geologists or paleontologists. The idea that the earliest forms of life may have existed along seashores where organisms are least likely to be preserved does not give promise of being helpful. If such were true, the numbers of fossil organisms might be reduced, but there still would be opportunity for the preservation of some of these ancestral forms.

Other paleontologists suggest that hard parts of these organisms did not develop until the Cambrian time. This view requires an *explosive* development of chitinous, calcareous, and siliceous hard parts at the same time, since exoskeletons of Cambrian animals are not limited to calcium for strength and hardness. This is replacing one problem with another.

Several paleontologists have proposed that *explosive evolution* occurred in the late Precambrian or early Cambrian period and that basic kinds developed quickly and have remained essentially unchanged in the 600 million or more years since that time.[34] But a few million years of time for explosive evolution is sufficient time for some intermediate forms to be preserved, yet none have been found. Furthermore, the study of heredity has shown no way for major mutations or evolutionary jumps to occur. This concept of explosive evolution comes closer to the idea of creation. One can substitute "creation" for explosive evolution in their theory.

Chapter Notes

Chapter Abstract

An investigation of the major groups of Cambrian invertebrates shows that they are not primitive ancestors to existing invertebrates but are already complex anatomically and physiologically. Those major groups found in sediment later than the Cambrian period appear abruptly without predecessors.

chapter four

Major Groups of Invertebrates

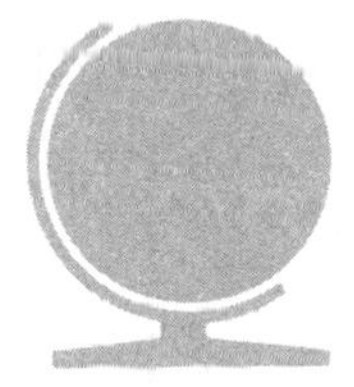

The major groups of the invertebrates appear in the Cambrian era and continue as unbroken lines to the present. Subdivisions within the phyla, such as classes and orders, appear abruptly in the Cambrian and later periods. These remain distinct and do not merge with representatives of other basic "kinds."

The phylogenetic (evolutionary) trees that are often seen in biology and paleontology textbooks give an impression that cannot be supported by the fossil record. Such trees are an exercise in speculation, as noted by M. T. Ghiselin of the University of California: "It is true that many works on phylogeny do read like imaginative literature rather than science. A disproportionate segment of the literature seeks to fill gaps in the data with speculations and nothing more."[35] Such diagrams (see figure 2) often show the branches and trunk of the tree as hypothetical, but the average person thinks that such trees do accurately represent past evolutionary history. However, it is well known that only the ends of the branches and the twigs are found as fossils. The limbs and trunk have yet to be discovered. To carry the analogy to a truly representative conclusion, the evolutionary tree should collapse because its major supports are missing. It should be replaced by a series of parallel lines, the main lines extending without branching. Instead of a widely branching evolutionary tree, a forest of individual trees best represents the past history of living things.

It might be useful to take a more detailed look at the fossil types that should be found, based on the predictions of the various models of origins. Then we can

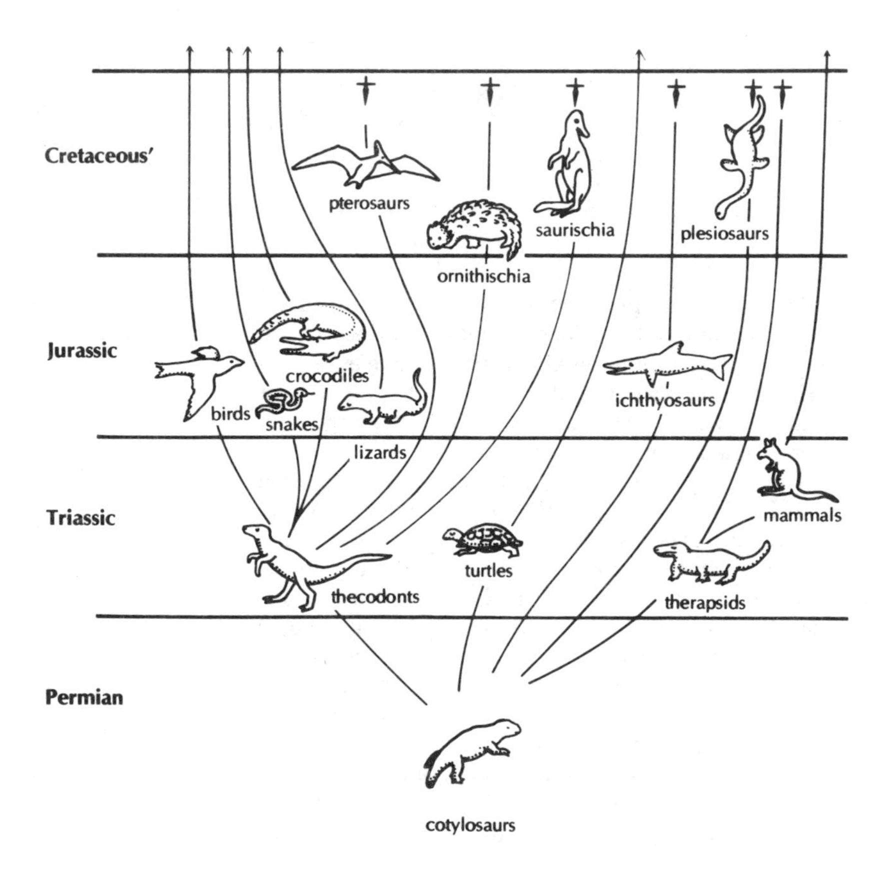

Figure 2. A typical phylogenetic tree for the fossil record of reptiles.

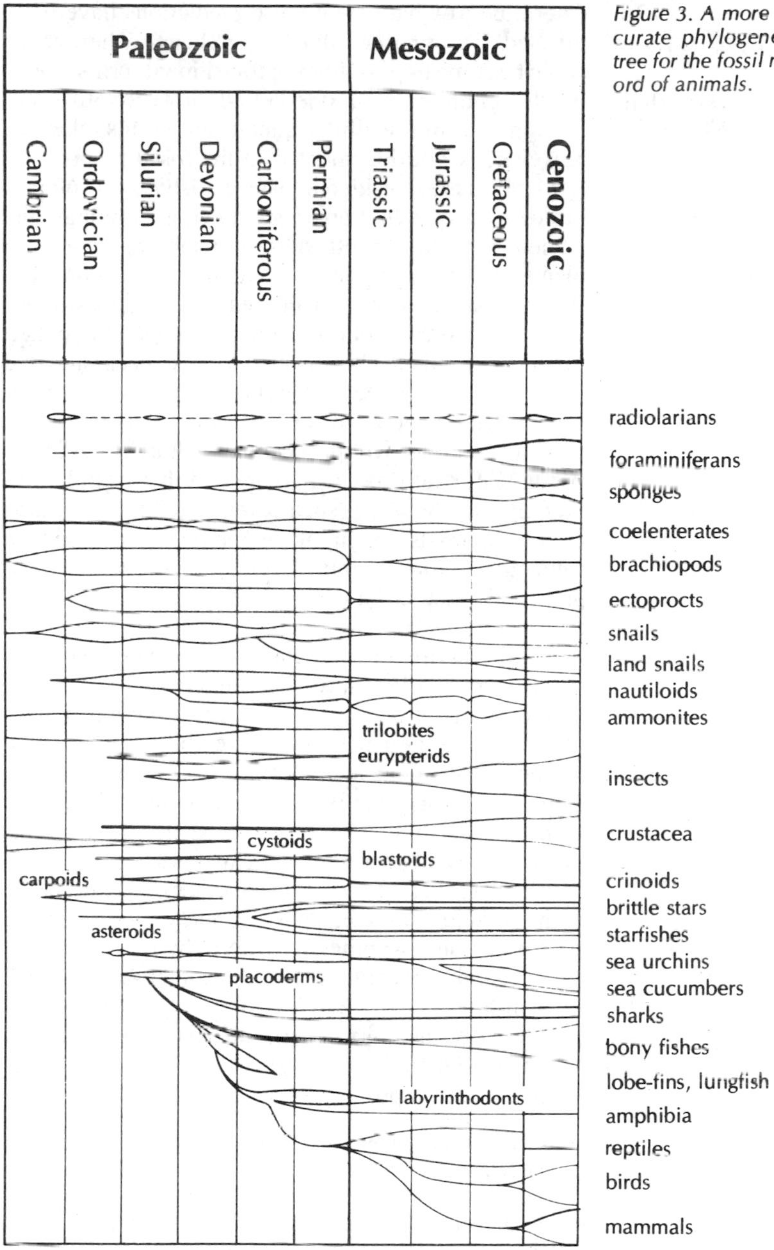

Figure 3. A more accurate phylogenetic tree for the fossil record of animals.

Paleozoic
Mesozoic
Cenozoic

Cambrian
Ordovician
Silurian
Devonian
Carboniferous
Permian
Triassic
Jurassic
Cretaceous

radiolarians
foraminiferans
sponges
coelenterates
brachiopods
ectoprocts
snails
land snails
nautiloids
ammonites
trilobites
eurypterids
insects
crustacea
cystoids
blastoids
carpoids
crinoids
brittle stars
asteroids
starfishes
sea urchins
placoderms
sea cucumbers
sharks
bony fishes
lobe-fins, lungfish
labyrinthodonts
amphibia
reptiles
birds
mammals

One Cell to Many Cells

determine to what extent these predictions have been fulfilled. The first possible major change where there could be some hope of finding transitional forms would be the change from one-celled protozoanlike organisms to multicellular metazoans. This change, however, is not documented in the fossil record.

Metazoa are considered to have arisen from colonial protozoa, a syncytial organism (having more than one nucleus but without cell walls or membranes), or from plants, depending upon which authority one consults. But there are those who dispute each of these positions.

It is easy to think of protozoans as *simple* one-celled organisms. This itself is an erroneous concept and one that must be avoided. No living organism is simple. Many of the protozoans are exceedingly complex; in fact, as complex within their one cell as some metazoans. Colonial protozoa compound that complexity. Attempts to show that such colonies led to true multicellular animals are highly theoretical and based on subjective evidence only. Even the opposite has been proposed: that colonial protozoans have developed from metazoa.[36]

The possibility of the metazoans being derived from plantlike protozoans has been attractive to some.[37] It would appear to be a simple matter for these primitive plants to lose their photosynthetic ability and thus be forced to obtain nutrients as animals do. However, the step from protozoanlike plants that have lost their chlorophyll, to metazoans with such structures as nerve tissue and an alimentary canal is a big one. Why would these metazoans evolve another highly complex form of nutrition when they already possessed the process of photosynthesis? Chlorophyll is a complex product and is considered by others to have developed later in time.[38]

In view of this maze of speculations and proposals, this statement by G. A. Kerkut of Southampton University seems appropriate:

> What conclusion then can be drawn concerning the possible relationship between the Protozoa and Metazoa? The only thing that is certain is that at present we do not know this relationship. Almost every possible relationship has been suggested [as well as many impossible ones], but the

> information available to us is insufficient to allow us to come to any specific conclusion regarding the relationship. We can, if we like, *believe* that one or other of the various theories is the more correct but we have no real evidence.[39]

In his book *Implications of Evolution,* Kerkut discusses which metazoans are the most primitive. He lists five: the Porifera (sponges), the Mesozoa (a small group of parasites), the Coelenterata (jellyfish, corals), the Ctenophora (sea walnuts), and the Platyhelminthes (flatworms). He indicates the primitive and advanced features of each group and concludes that it is impossible to know which of these is the most primitive metazoan.[40] As we search for the primitive forms that should bridge the gap between protozoans and metazoans and as we attempt to identify the most "primitive" metazoans, we are frustrated in our attempts. We lack good historical evidence for any change from one cell to many cells. Each proposal is speculative, and we cannot come to any valid scientific conclusion about this postulated transition between protozoans and metazoans.

Cambrian Invertebrates: Primitive or Advanced?

The terms *primitive* and *advanced* are frequently found in discussions of evolutionary development and phylogenetic relationships. Is it possible to look for and add up certain features in various organisms and decide which one is most primitive? Such a procedure would make the study of evolution relatively simple. Unfortunately, the designations *primitive* and *advanced* are subjective and are based on the opinions of researchers who assume that evolution has taken place. If they assume that evolution has moved in certain directions, they are preprogramed to think that certain features are more primitive and represent earlier stages of development. But the circular nature of this type of reasoning becomes apparent. If an individual does not assume that the evolutionary models are true, he is hard put to identify what is "primitive" and what is "advanced." Characteristics that have been considered by one researcher as "primitive" frequently have been identified by another as more specialized and "advanced."

The earliest multicellular organisms found in the Cambrian period are brachiopods (lampshells), archeocyathids, arthropods (trilobites and crustaceans), and annelids (segmented worms). Brachiopods are small soft-bodied marine animals with an upper and lower shell. They were originally classified with the mollusks. However, the structure of the soft parts of brachiopods is different from that of the mollusks, and, as a result, they have been assigned to a distinct phylum. The archeocyathids are an interesting group of corallike or spongelike animals that appear only in the lower or middle Cambrian formations. No ancestors or descendants are known. Their relationship to other invertebrates is not clear, but they are now considered to be in a phylum by themselves.

Arthropods, especially trilobites, are among the earliest fossils. Annelid worms are not abundant, but their trace fossils (i.e., burrow marks) are seen regularly in early Cambrian sediments and also have been reported occasionally in the Precambrian. What is the nature of these early Cambrian fossils? Are they primitive, early stages in the evolution of worms, trilobites, and lampshells, or are they fully developed? An investigation of these Cambrian fossils should answer these questions.

Although fossils seldom preserve traces of soft parts, there are some exceptions. The Burgess Shale specimens found in the Canadian Rockies in fine-grained shales show many of the internal organs. But even in fossils that do not show soft parts, much can be learned concerning the physiology of the animal. Evidences of eyes and antennae give us information about the nervous system; gills and other respiratory structures tell us something about the respiratory system; and the mouth parts indicate the kind of food that the animal obtained from its environment.

Thus, step by step, it is possible to build a fairly complete understanding of the physiology of animals for which we have only the preservation of the hard parts. Figure 4 describes the external anatomy and makes deductions concerning the physiology of animals in two phyla: the Arthropoda (arthropods) and the Annelida (worms). An examination of this table re-

Type of Animal	Visible Morphological Characteristics	Implications Based on Visible Characteristics
TRILOBITES (Arthropoda)	1. Chitinous exoskeleton	Growth by periodic ecdysis (molting), which is a very complex process
	2. Segmented body	
	3. Jointed legs and appendages	Complex musculature
	4. Compound eyes	Complex nervous system
	5. Antennae	
	6. Bristles for swimming, and respiration	Blood circulatory system
	7. Complex mouthparts	Specialized food requirements
	8. Great variation	
SEGMENTED MARINE WORMS (Annelida)	1. Segmentation	Repetition of certain organs in each segment
	2. Complete digestive system (mouth→anus)	
	3. Setae and bristles	Complex musculature
	4. Chitinous jaws .... (definitely found in Ordovician and probably also in Cambrian)	Specialized food habits and requirements; complex musculature
	5. Eversible proboscis...	
	6. Eyes..........................	Nervous system similar to living segmented worms
	7. Respiratory filaments and gills	Blood circulatory system
	8. Great variation	

Figure 4. Complexity of Cambrian representatives of two animal phyla

veals that in no way can trilobites and annelids, even from the Cambrian period, be considered simple or primitive. Modern counterparts that live in the oceans today are no more complex. The same type of comparison could be made with fossil and modern brachiopods, but not with archeocyathids, since they are extinct. Not much is known in this case about the nature of the soft tissue of the animal.

The Absence of Invertebrate Ancestors

Many animals are so different from any other type of organism that their evolution is difficult to imagine. Radiolarians are exquisite little one-celled creatures that usually have an external siliceous skeleton that is ornately constructed and perforated. They are often abundant in marine sediments and have been reported as numerous in Precambrian rocks of France.[41] Ancestors are unknown and the hope of finding them in lower rocks is remote.

An abundance of trilobite species appear in the lowest beds of the Cambrian period according to Norman Newell:

> Several superfamilies comprising many families appear abruptly near the base of the Cambrian; others are introduced somewhat higher in the Lower Cambrian. Of the ten superfamilies of Lower Cambrian trilobites, not a single ancestor is known.[42]

Other arthropods also are distinct and appear abruptly. In his book *Prehistoric Life,* Percy Raymond mentions several:

> Spider-like animals have been found in the Lower Carboniferous, and true spiders as well as various related forms are known from the Pennsylvanian. Nothing is yet known of the ancestry of these groups. . . . Specimens found in the Rhyme chert of Scotland show that mite-like creatures were in existence as early as Mid-Devonian times, but their aquatic ancestors have not yet been discovered. . . . It is somewhat exasperating that so little is known of their origin.[43]

The origin of the insects is no better known even though they are the most abundant animals in existence today.

> The presence of several hundred species of insects in the Pennsylvanian makes their sudden appearance at this time the more remarkable. The diversity of the forms represented implies a long antecedent evolution whose record may yet be found in Mississippian if not in Devonian rocks.[44]

Judging from past experience, the hope expressed above is not likely to be fulfilled. The creation model would predict that we would not find the postulated ancestors because they never existed.

Echinoderms are a distinctive group of invertebrates. Speaking especially of the crinoids (sea lilies), James Beerbower says, "All are distinct at their first appearance. Indeed, the connections between orders in the same subclass cannot be established with certainty."[45]

The ammonites (figure 5) are an extinct subclass of mollusks that constitute important fossils. They usually are coiled like giant snails but are considered to be related more closely to the squid, octopus, and chambered nautilus. A. H. Clark writes:

> Most interesting in this connection are the ammonites, for they begin as ammonites, become enormously diversified, then disappear without ever being anything but ammonites. Here we have a group . . . of which the entire history seems to be laid bare for our inspection. The early portion of the history of ammonites is not lost in the unknown pre-Cambrian.[46]

Thus, in the case of the ammonites, we have their entire life history from origin to extinction. And yet we do not see any evidence for the models of evolution.

Annelids and arthropods are considered to be related evolutionarily through a common ancestor. One of the creatures proposed as a transitional form between the two groups is the caterpillarlike worm *(Peripatus)* belonging to the subphylum Onychophora. Peripatus (figure 6) does have a number of wormlike features such as segmentation, no true-jointed legs, and a wormlike body wall. Its arthropodlike characteristics include an insectlike blood system, and respiratory organs consisting of tracheae and spiracles. There are also characteristics that are unlike those of either ar-

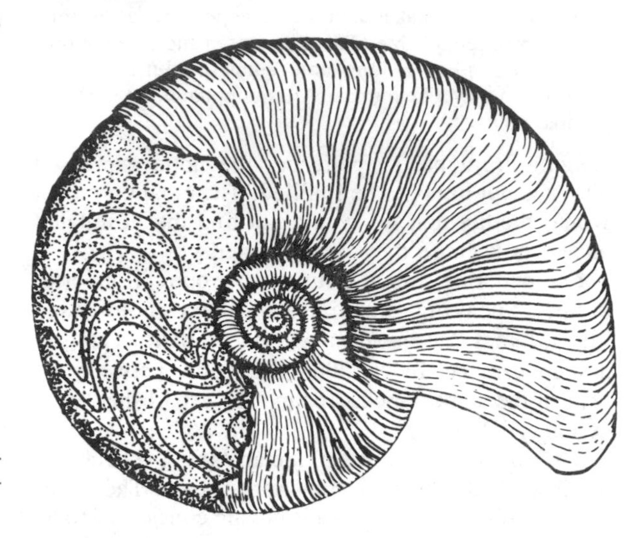

Figure 5. Shell structure of a fossil ammonite.

thropods or annelids. Onychophoran worms of the genus *Asheaia* (figure 6) have been identified in the Burgess Shale of the Middle Cambrian.[47] However, since both annelids and arthropods are abundant in the Cambrian period, the evolution from annelid to arthropod must have occurred, if it did so at all, either in the very early Cambrian or in the Precambrian. As has already been mentioned, fossil evidences are rare in the Precambrian sediments; consequently the chances of finding the ancestors of all three (onychophorans, annelids, and arthropods) are extremely slim. The fact that we find all three in Cambrian rocks, fully developed, and without ancestors in the Precambrian, leads us to believe that there never have been any forms transitional between these three groups.

Another specimen from the Burgess Shale is revealing. An arrowworm, phylum Chaetognatha, was identified.[48] This places another distinct major category back into the Cambrian. It is not necessary to marshal more examples. The fossil record is singularly silent on the origins of the main groups of invertebrate animals. This statement by A. H. Clark, though written

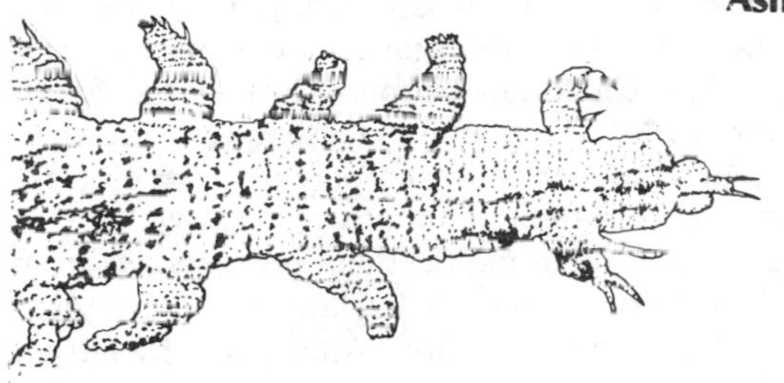

Peripatus

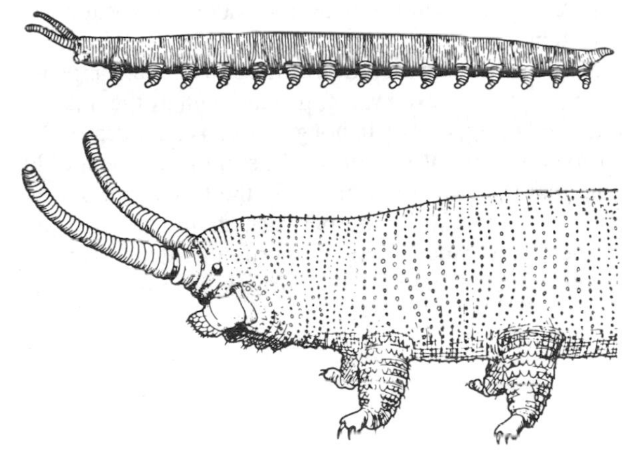

Figure 6. Comparison of a Cambrian onychophoran worm ASHEAIA *with a modern onychophoran worm* PERIPATUS *(overview and front segment).*

years ago, summarizes our present position in paleontology:

> Since all the fossils are determinable as members of their respective groups by the application of definitions of those groups drawn up from and based entirely on living types, and since none of these definitions of the phyla or major groups of animals need be in any way altered or expanded to include the fossils, it naturally follows that throughout the fossil record these major groups have remained essentially unchanged. This means that the interrelationships between them likewise have remained unchanged.[49]

In conclusion, what can we say about our examination of the invertebrate fossil record? First, we find that there is a sudden appearance of complex invertebrates in the Cambrian period. Most of the major invertebrate phyla are found in these rocks, and when we attempt to trace the ancestry of these invertebrates, they are nowhere to be found. Second, as we seek to trace the evolution of multicellular invertebrates from one-celled animals, we are lacking solid historical evidence for such a transition. Attempts to identify a particular invertebrate group as most primitive have proved futile.

We have also seen that those invertebrates that first appear in the Cambrian period have been complex invertebrates. In no way can these early multicellular organisms be considered primitive. The later appearance of other invertebrates, groups such as the ammonites and insects, also do not give us any support for the various models of evolution. In summary, we should note that the explosion of life in the Cambrian period and the systematic gaps between major invertebrate "kinds" are much more supportive evidence for the creation model than for the various models of evolution.

Chapter Notes

Chapter Abstract

An analysis is made of the vertebrate fossil record. Evidence for the origin and subsequent evolution of the vertebrates is evaluated. The claims for vertebrate transitional forms are reconsidered.

chapter five

The Vertebrate Fossil Record

Let us now turn our attention to the fossil record of the vertebrates. This group of animals should provide us with an abundance of material that can be used to test the validity of our various models of origins. We should expect to find good documentation, since vertebrates are generally much larger than most invertebrates and they tend to leave behind conspicuous and diagnostic material for paleontologists to analyze (skulls, appendages, and vertebrae).

The Origin of the Vertebrates

The vertebrates are assumed by evolutionists to have evolved from invertebrate ancestors in the Cambrian or Precambrian periods. The transition from an invertebrate to a vertebrate is so great that researchers have been much perplexed concerning this aspect of evolution. Nearly every major group of invertebrates has been proposed as a potential ancestor for the vertebrates.

The most popular view is that a multicellular larvalike animal (planula) evolved into a simple flatworm, which in turn evolved into echinoderms, certain worms, and chordates (vertebrates). It must be noted that there is a great difference between invertebrate echinoderms and the vertebrate chordates.

The main evidence in support of this evolutionary scheme comes from the similarity of the larvae of echinoderms to those of hemichordates. The *Hemichordata* (certain sea worms) are said to have traces of chordatelike characteristics, thus serving as a link from echinoderms to vertebrates. This tenuous connection of echinoderms with vertebrates cannot be fairly entertained without also considering the great differences between them.

First, echinoderms, unlike the vertebrates and many of the other invertebrates (arthropods and mollusks), do not have any head structure. Second, echinoderms lack any semblance of a nerve center or brain, such as is found in the vertebrates. Third, echinoderms lack much of the internal physiology found in vertebrates. They have a rudimentary and open circulatory system, and they do not have an excretory system. Finally, echinoderms possess a unique water vascular system that functions as a hydraulic system to aid the animal in locomotion, feeding, and respiration.

Some biochemical links between vertebrates and echinoderms mentioned in the scientific literature are not soundly established. In fact, further data have negated the evolutionary inferences previously drawn from incomplete information.[50]

Another evolutionary sequence from one-celled organisms to vertebrates has been proposed.[51] This scheme is much different from the ones described above, since it postulates this transition through a flukelike creature and a segmented worm, to the first vertebrates. For the segmented ancestor of the worms and arthropods to develop into vertebrates, it would be necessary for the blood flow to reverse directions. In worms, the blood vessel along the top transports blood toward the front, and the bottom vessel carries blood away from the heart. Vertebrate blood, however, flows in directions opposite to this. Also, the ventral nervous system of annelids would have to move to a dorsal position above the alimentary canal and change from a solid to a tubular structure. Further, the anal opening would have to shift from the posterior end to a position farther forward.[52]

When we turn to the fossil record, no evidences of any of the proposed transitions from invertebrates to vertebrates can be found. Both the Cambrian and Precambrian periods have been suggested as the time frame of this alleged transition. Some investigators believe they have found the trace fossil evidence scales for vertebrates in Cambrian rock.[53] If we accept this evidence, we lack any possible ancestors in the Precambrian rocks. If we accept the traditional evolutionary scheme that postulates an origin of the vertebrates during the Cambrian period and their first appearance in the Ordovician period, we still are without ancestors, because none have been found there.

> How this earliest chordate stock evolved, what stages of development it went through to eventually give rise to truly fishlike creatures, we do not know. Between the Cambrian, when it probably originated, and the Ordovician, when the first fossils of animals with really fishlike characteristics appeared, there is a gap of perhaps 100 million years which we will probably never be able to fill.[54]

One might rightly ask, is there any evidence of this transition, or is all of this merely speculation? This surprising quote from N. J. Berril's book *The Origin of Vertebrates* might answer this question:

> There is no direct proof or evidence that any of the suggested events or changes ever took place; what strength the argument may have comes from whatever wealth of circumstantial detail I have been able to muster. In a sense this account is science fiction, but I have myself found it an interesting and enjoyable venture to speculate concerning the Cambrian and Precambrian happenings that may have led to my own existence.[55]

The Fossil Record of the Fishes

Between each of the major classes of fishes (Agnatha, Placodermi, Chondricthyes, and Osteichthyes), there are consistent gaps that are not bridged by any transitional forms. The jawless fishes (class Agnatha) initially appear in the Ordovician period without predecessors. They are followed by the abrupt appearance of the other major orders of Agnatha in the Silurian period.[56]

The placoderms are a group of extinct fishes. Each of the major types of placoderms appears at the Silurian-Devonian border and is not preceded by any transitional forms.[57] The placoderms do not fit a scheme of progressive evolution.

> We would expect "generalized" forms that would fit neatly into our preconceived evolutionary picture. Do we get them in the placoderms? Not at all. Instead, we find a series of wildly impossible types which do not fit into any proper pattern; which do not, at first sight, seem to come from any possible source or to be appropriate ancestors to any later or more advanced types. In fact, one tends to feel that the presence of these placoderms, making up such an important part of the Devonian fish story, is an incongruous episode; it would have simplified the situation if they had never existed![58]

The origin of the class Chondrichthyes (the sharklike, cartilaginous fish) also poses a problem for the traditional scheme of progressive evolution because it does not fit into any expected pattern.[59] In fact, the original theories for the origin of this group based on evidence from embryology have been reversed because of the fossil evidence. The elasmobranch fishes (sharks and rays) were originally considered to be "primitive" fish. However, it turns out they are actually one of the last groups to appear in the fossil record. Thus, the evidence from paleontology has forced a reversal of the original belief that these fishes were primitive.[60]

The origin of the bony fishes (the Osteichthyes) is also quite speculative. Their appearance in the fossil record is a "dramatically sudden one" and a "common ancestor of the bony-fish groups is unknown."[61] Within this group of bony fishes there are also consistent gaps between major groups that are not bridged by transitional forms.

Fin Today, Foot Tomorrow

The evolutionary model postulates that representatives of the rhipidistian fishes gave rise to the first land tetrapods, the amphibians. Some evolutionists suggest a different group of fishes gave rise to each different group of amphibians[62] and others suggest that one group of fishes gave rise to all of the different groups of amphibia. [63] The fact that two competing theories can be presented to explain the origins of amphibians suggests that there is a substantial lack in our information about this transition. The evidence is insufficient to document either theory.

The first amphibian to appear in the fossil record is *Ichthyostega* (see figure 7). Its appearance in the Devonian period in Greenland is abrupt and without transitional forms. *Ichthyostega* is already quite specialized by the time we see him in the fossil record and can be considered quite advanced.[64] The supposed evolutionary ancestor is the rhipidistian fish *Eusthenopteron* (see figure 7). This amphibian exhibits certain specializations that make it exclusively aquatic.

The difference between an aquatic mode of life and a semiterrestrial mode of life are considerable. Life in shallow water would require pelvic and pectoral girdles, a flexible neck, and sensory apparatus (sight and hearing) that would be functional out of the water.[65] Further modifications in anatomy (ossification of vertebrae, development of limbs and digits) and physiology (respiration, excretion) would be necessary for the

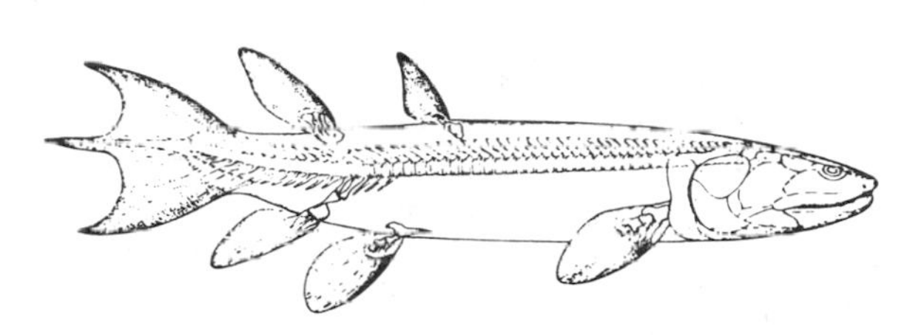

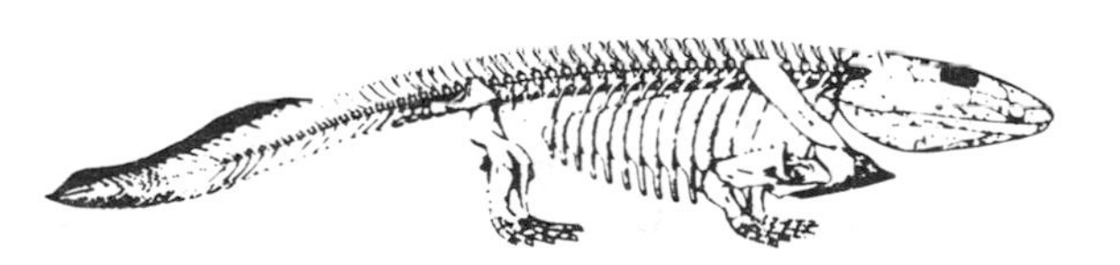

Figure 7. Comparison of the rhipidistian fish EUSTHENOPTERON *(top) with the amphibian* ICHTHYOSTEGA *(bottom).*

final transition we find in *Ichthyostega*.[66] But when we turn to the fossil record for evidence of this important transition, we find none.

An important feature in this postulated transition from water to land is the development of a foot from a fin. Early authors suggested that the digits and adjacent bones of the foot were new structures,[67] but the current view is that these limb bones are rearranged fin bones.[68] The problem with this view is that there is no fossil evidence to support it.

The various models of evolution also require that the process of natural selection provide a mechanism for the change from fish fins to amphibian feet. Evolutionists have engaged in a great deal of speculation in an attempt to identify some environmental cue that would provide a force sufficient to evolve feet from fins. It has been suggested that these limbs aided in migration during drought conditions or that these limbs were responsible for both migration and food capture.[69] It has also been suggested that these limbs would have provided support that would be necessary for feeding near the water's edge[70] or that these limbs would allow for burrowing during drought conditions.[71] Other more recent theories suggest that migration is the major selective factor that led to the evolution of these limbs. However, the evolutionists who have proposed these theories suggest that this migration pressure was due to a need to decrease population pressure,[72] to decrease larval competition,[73] or to increase the range of these species.[74]

Each of these theories attempts to explain why fish evolved into amphibians. It is difficult to imagine how any factor would be sufficient to provide selection for the origin of complex terrestrial appendages. It requires a high degree of faith to accept one of these proposals. It is faith based, *not* on objective fossil evidence, but rather on the assumption that evolution has indeed occurred.

Within the class of amphibians, we also find consistent gaps between major taxonomic groups. For example, there are three orders of amphibians living today. Each of these groups appears in the fossil record without transitional forms. The ancestors of these groups should be found in late Devonian or Mississippian rocks, but none have been discovered. Furthermore, between these periods and the first appearance of one of these groups, there is, according to Alfred S. Romer, "a broad evolutionary gap not bridged by fossil materials."[75] This first appearance of fossils that would be considered to be frog representatives occurs in the Triassic period. This fossil is modern in almost every respect,[76] and thus sheds no light on the ancestry of the frogs.[77] The first members of the two other orders appear even later and do not provide any conclusive evidence for the ancestry of these groups.

Thus, as we examine the postulated origin and evolutionary history of the amphibians, we do not find evidence for the models of evolution. We lack transitional forms to bridge the gap between aquatic fish and semiterrestrial amphibians with feet. We find also that the modern amphibians appear in the fossil record without transitions and are essentially modern in appearance.

The Skin-to-scale Transition

If amphibians evolved into reptiles, the fossil record should provide documentation for this important event. *Seymouria* has been suggested as one possible transitional form (see figure 8). "*Seymouria* is a small tetrapod found in the upper portion of the lower Permian sediments exposed to the north of the town of Seymour, Texas."[78] The major difficulty with *Seymouria* is that the period in which it is found is too late in the rock record to be the grandparent of the reptiles. It is found in the Permian period, while reptiles such as *Hylonomus* (figure 8), are found as early as the Pennsylvanian period.[79] Thus, we have a situation in which the children are older than the grandparents. Further investigation has shown that *Seymouria* probably had an aquatic amphibian life history and should be classified with the amphibians.

Seymouria was selected as a transitional form because of the number of reptilian skeletal features it possessed. Unfortunately, skeletal features are not always a good indication of an organism's relationship with other organisms. Modern amphibians can easily be distinguished from modern reptiles by means of their external soft parts. The fossil record, however, does not preserve soft parts, thus it is often impossible to make a clear distinction. Furthermore, reptiles differ from amphibians in that reptiles lay an amniotic egg. Although this is a good diagnostic feature, it is not

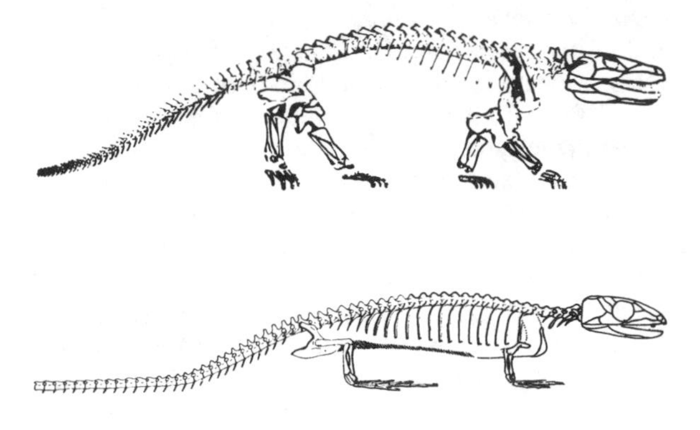

Figure 8. SEYMOURIA *(top), a proposed transitional form between the amphibians and reptiles and* HYLONOMUS *(bottom), the first reptile found in the fossil record.*

preserved with any regularity in the fossil record. In summary, it should be noted that our difficulty in classifying certain fossil finds of amphibians and reptiles is not a result of a close affinity between the groups, but it is rather a result of a fossil record that cannot give us all of the facts we need to make an objective decision.

The Birds: Born Airborne?

Between the reptiles and the birds there is a gap in anatomy and physiology. In that gap stands one famous bird considered by evolutionists to be the best example of a transitional form: *Archaeopteryx* (figure 9). This unusual creature is known solely from the Solenhofen limestones in Germany. *Archaeopteryx* is considered to be a transitional form because it possesses a number of features that have been interpreted as reptilian.[80] However, it also possesses a number of avian (birdlike) features. Thus, this creature is what would be predicted by the models of evolution. The question that should be raised is this: Can the creation model also account for this creature, or is it an anomaly that contradicts the creation model?

First, it should be noted that some of its characteristics that are considered reptilian can also be found in other living and extinct birds. For example, *Archaeopteryx* had claws on its wings. However, many other birds have claws on their wings as well.[81] Many birds have a claw on the first digit (finger) and gallinaceous birds (grouse, turkey, chicken), birds of prey, waders, and swimming birds have a claw on the second digit in the embryo and occasionally in the adult. Many ratites, such as the rhea *(Rhea)* and the ostrich *(Struthio)*, possess claws on the first, second, and third digits.[82]

There is also a bird found in South America known as the hoatzin *(Opisthocomus houzin)*, which possesses strong claws on its first and second digits and has rudimentary claws on its third and fourth digits.[83] Thus, it would be considered more primitive than *Archaeopteryx* because of its claws and other anatomical features.

Another feature of *Archaeopteryx* that is considered reptilian is its teeth. Modern birds do not possess teeth, but we know that some extinct birds had teeth. Reptiles of that period had a wide variety of dentition, while most modern reptiles do not have true teeth. The teeth of *Archaeopteryx* do not fit nicely into any scheme of progressive evolution, and they do not seem to be good candidates for transitional structures.[84]

Second, we should consider what ecological niche this creature might occupy. In other words, what types of food did *Archaeopteryx* eat and how did it obtain it? It is probable that it was a carnivorous (meat-eating) predator. Thus, we would expect that it would have features that would make it more successful as a carnivorous predatory bird. This is precisely what we find with *Archaeopteryx*. It had a number of anatomical features that were similar to those found in predatory ground birds and small predatory mammals. It may have been a predator that found good use for claws on its wings and teeth in its jaws.

As was discussed earlier, transitional structures do not necessarily imply evolutionary relationships. If we, however, assume that *Archaeopteryx* is an evolutionary transitional form between the birds and reptiles, we would like to identify its ancestors. Origi-

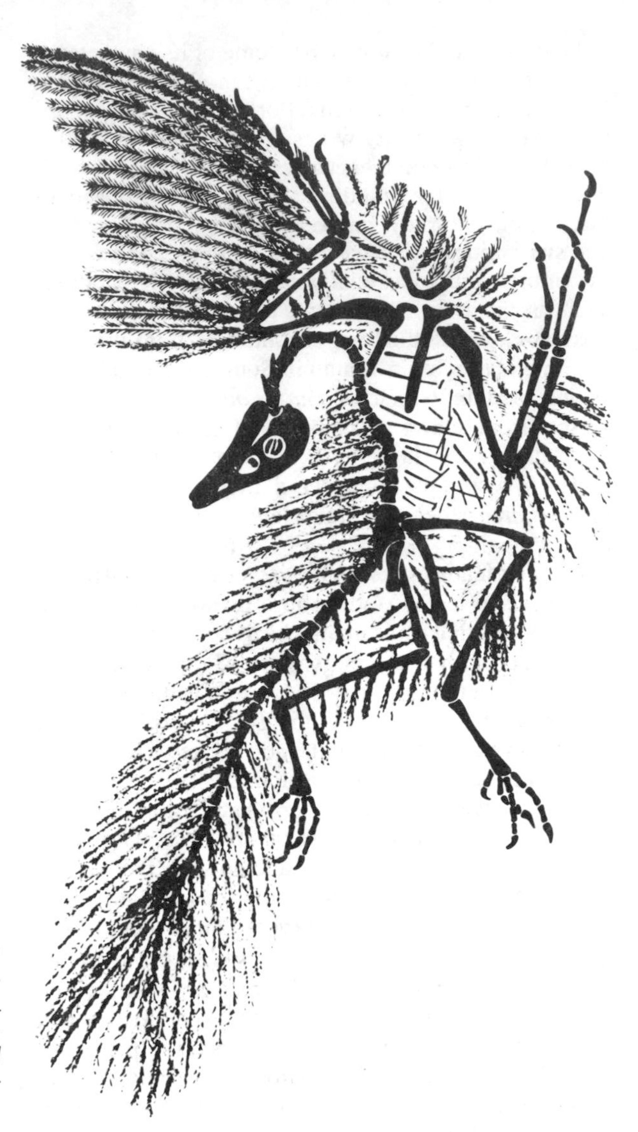

Figure 9. A fossil skeleton with feather impressions of ARCHAEOPTERYX, *a proposed transitional form between reptiles and birds.*

nally, *Archaeopteryx* was thought to have evolved from thecodont reptiles,[85] but a more recent theory proposed by John Ostrom of Yale University suggests that *Archaeopteryx* shared a common ancestor with the theropod dinosaurs.[86] The fossil record does not, however, document this transition, or any other that has been proposed.

The most important diagnostic features of *Archaeopteryx* are those that are avian. *Archaeopteryx* is considered to be a bird because it has feathers and wings. These feather impressions are not primitive feathers, nor are they transitional structures between scales and wings, but they are modern feathers in every respect. The presence of feathers also implies that *Archaeopteryx* possessed the internal physiology (a warm-blooded circulation) and biochemistry (an endocrine system for molting of feathers) that are generally associated with modern birds. The presence of wings with feathers is sufficient to classify it with all modern birds.[87]

The bone structure of *Archaeopteryx*, although different from typical birds, is certainly avian. "Aside from the feather impressions, the most significant avian feature of *Archaeopteryx* is the furcula or wish bone. . . . Its presence, together with feathers, confirms the avian identification of *Archaeopteryx*."[88]

In conclusion, it must be stressed that the transition from reptiles to birds is still a hypothetical construction. According to W. E. Swinton of the British Museum of Natural History in London, "the origin of birds is largely a matter of deduction. There is no fossil evidence of the stages through which the remarkable change from reptile to bird was achieved."[89] It is important to note that Swinton did not say that there are only a few fossils at this supposed transitional stage. Instead, he says that there is *no* fossil evidence of this stage. This is what we would predict from the creation model.

The Scale-to-Fur Transition

The search for reptilian-mammalian transitional forms presents a very difficult problem. Most of the characteristics that differentiate a reptile from a mammal are based either on physiology (warm-bloodedness) or on parts of their anatomy that are not identifiable in fossil material (hair, milk). Attempts to correlate these physiological and anatomical changes with changes in skull and skeletal arrangement have created difficulties in the area of taxonomy. It is assumed that all mammals can ultimately trace their origins back to a single common ancestor. However, if we trace the changes in skull and skeletal material back to this postulated origin, it seems to suggest a polyphyletic origin (origin of many groups simultaneously) of the mammals. One way in which evolutionists have attempted to solve this problem of taxonomy is by classifying certain extinct creatures that we know are definitely reptiles, as mammals.[90] Again, as with the reptiles and amphibians, our problem in tracing the ancestry of mammals is a result of the inadequacies of the fossil record.

Most of the potential transitional forms are constructed from fragmentary fossil material consisting primarily of teeth and jaws. Proposed transitional forms are questionable and often appear too early in the fossil sequence to be related to their supposed reptilian ancestors.

We find each of the major orders within the class of mammals appearing abruptly in the fossil record. Many of these orders are considered to have evolved from an ancestral insectivore (a shrewlike animal), since these insectivores are the first potential modern mammals to appear in the fossil record. The relationship of these potential ancestors to the rest of the mammals is not based on a long sequence of transitional forms bridging these groups, but it is based almost entirely on subjective and controversial evidence from comparative anatomy and physiology.

The lack of transitional forms here is not an isolated case; this problem arises throughout the animal kingdom. George Gaylord Simpson has noted that: "This regular absence of transitional forms is not confined to mammals, but is an almost universal phenomenon, as has long been noted by paleontologists."[91] As we have seen so far in this book, the regular absence of transitional forms is indeed an almost universal phenomenon. There is an abrupt appearance of most of the major invertebrate and vertebrate phyla, classes, and orders throughout the geological column. As we look for ancestors and transitional forms, we find very few. Those which have been identified as potential transitional forms have been shown to be reconcilable with a model of creation. Therefore, we can conclude that the model that best explains the data from the fossil record is the creation model.

Chapter Abstract

If the models of evolution are true, the fossil record should especially document the evolution of new structures. The origin of flight and the adaptations necessary for aquatic locomotion are considered.

chapter six

The Origin of Adaptive Structures

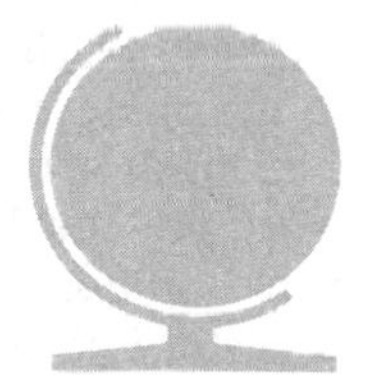

Other tests could be applied to the fossil record of animals that would help to determine the validity of the various models for origins. For example, evolutionary models must postulate a long and intricate series of transitions for an animal in one environment to adapt to a totally different environment.

In many cases, these adaptations would require a complex series of anatomical changes. According to evolutionists, life arose in the water. Then, independently, both vertebrates and invertebrates made their assault onto the land. After achieving some success, these animals gave rise to others that successfully made other transitions from life on land to life in the air and to life back into the water.

If such an evolutionary history took place, we would expect to find occasional fossils that would document such a transition. However, if the creation model is correct, we would predict that there would be an absence of transitional forms between these fossils, since they represent created and distinct "kinds" of organisms.

The Origin of Flight

One of the most crucial tests for these models can be found in the fossil record for the origin of flight. Flight supposedly evolved independently four times during geologic history. However, Everett C. Olson of the University of Chicago has said in his book *The Evolution of Life* that "as far as flight is concerned, there are some big gaps in the [fossil] record."[92]

Of the invertebrates, the insects are the major group to achieve the ability to fly. Although the rock record for insects is far from complete, we can identify many insects of the past. It does not, however, give us any information about the origin of wings. According to F. M. Carpenter of Harvard University, "the process by which wings were acquired by insects has been a question of much speculation, for they are not modifications of previously existing appendages."[93] In fact, "there is almost nothing to give any information about the history of origin of flight in insects."[94]

Within the vertebrates, there were three major groups that supposedly evolved flight. The first of these was the flying reptiles (figure 10). When the flying reptiles appear in the Jurassic period, they are fully capable of flight and "there is absolutely no sign of intermediate stages."[95]

The next group capable of flight was the birds. The birds are the most abundant of the land vertebrates today; thus, we might expect to find some trace of flightless bird ancestors. As we have noted, the fossil record does not provide a sequence of fossils that shows the origin of wings and feathers. The first bird in the fossil record is *Archaeopteryx* (figure 9) who already had well-developed wings and fully formed feathers.[96]

The last group of animals with the capacity of flight was the mammals. Of the mammals, only the bats successfully developed flight. The earliest fossil evidence of flying bats "is in fully developed bats of the Eocene epoch."[97] This first bat is *Icaronycteris index*. By comparing it with a modern brown bat *Myotis myotis,* we can see that it is quite well developed

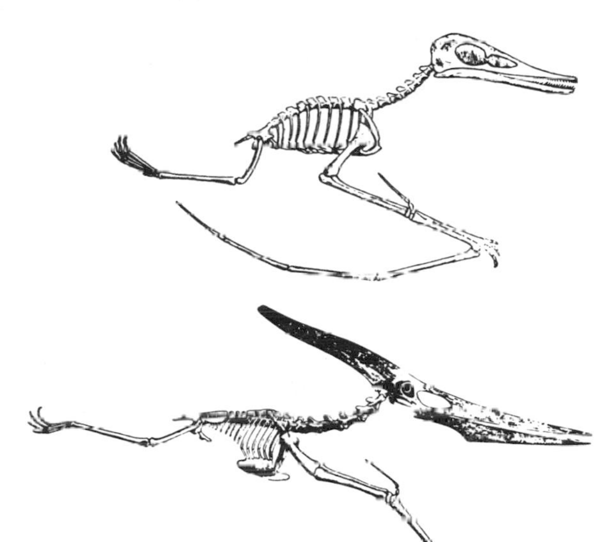

Figure 10. The bone structure of two of the flying reptiles— PTERODACTYLUS *(top) and* PTERANODON *(bottom).*

(figure 11). If bats evolved from shrewlike ancestors, there is certainly no fossil evidence to support it. *Icaronycteris* is a very advanced flying mammal.[98] Glenn Jepsen of Princeton University has stated that this Eocene bat "is not a 'missing link' between shrews or anything else and bats, but already a true bat."[99]

Back to the Water

Although only a few groups supposedly took to the air, there are a number of animal groups that evolutionists postulate returned to the water. It is not possible in this monograph to investigate every supposed transition. Therefore we will consider only a few cases from the vertebrates. If the models of evolution are true, the fossil record for the reptiles and mammals certainly should reveal some animals that made this change. But "if we seek the ancestries of the marine reptiles and mammals, we run into a blank wall as far as intermediate stages between land and sea are concerned except in the case of the plesiosaurs."[100] Addi-

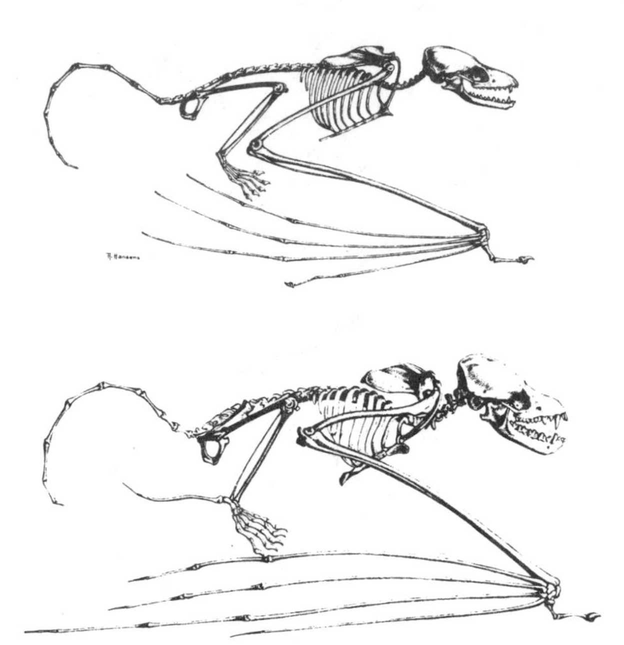

Figure 11. Comparison of the first bat ICARONYCTERIS INDEX *(top) with the modern bat* MYOTIS MYOTIS *(bottom).*

tional investigation has even shown that plesiosaurs are not preceded by transitional forms. Those forms are no longer considered to be intermediate stages between terrestrial reptiles and plesiosaurs.

Among the reptiles, there are a number of aquatic groups. Each of these appears as separate and distinct. The turtles (order Chelonia) appear abruptly and "little light is shed on the ultimate origin of the turtles from a study of fossil members of the order."[101] The first turtle to appear in the fossil record is already a true turtle and not a transitional form.

Turtles have a unique anatomy. Scales and dermal bones are greatly thickened and the rib cage is fused to these hard encasing structures. The transition from amphibians or other vertebrates that have their fore- and hind-limb girdles outside the rib cage to turtles that have these girdles inside the rib cage would be a difficult one requiring many intermediate steps. Some of these steps would appear as anatomical monstrosities! There are no fossils that even suggest such a change took place.

Another group of reptiles, the mesosaurs (order Mesosauria), were a fresh-water, fish-eating group. They were slimly built reptiles about a meter in length. Their origin is unknown and the earlier suggestion that they were ancestors to another reptilian group, the ichthyosaurs (figure 12), has been disproved by comparative studies of their bone structure.[102]

Other groups of aquatic reptiles can be found in the euryapsids (subclass Euryapsida). These are the nothosaurs, plesiosaurs (figure 12), and placodonts. The fossil record does not provide any ancestors for this group; so "we have no positive clues about the ancestry of the aquatic euryapsids as a whole."[103]

It was originally suggested that the nothosaurs, which were less specialized for aquatic life, were the ancestors to the plesiosaurs (figure 12). The nothosaurs are sufficiently different in anatomy from the plesiosaurs that paleontologist Alfred S. Romer has stated that nothosaurs "cannot have been plesiosaur ancestors."[104]

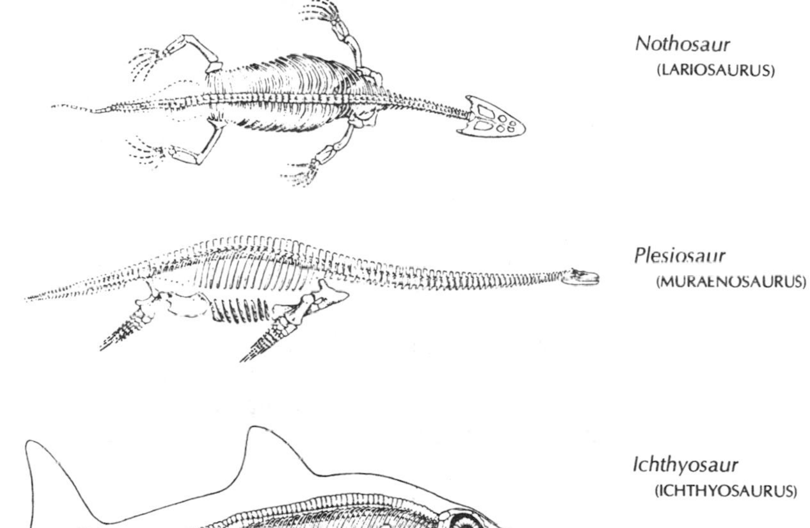

Nothosaur
(LARIOSAURUS)

Plesiosaur
(MURAENOSAURUS)

Ichthyosaur
(ICHTHYOSAURUS)

Figure 12. Bone structure of three groups of aquatic reptiles.

Finally, we come to the aquatic ichthyosaurs (see figure 12). These creatures resemble porpoises in appearance and were highly specialized for aquatic life. Their streamlined appearance and the numerous digits in their flippers should be found in other reptilian groups. If these ichthyosaurs have an evolutionary history, we would expect to find suitable ancestors leading through a series of transitional forms to these ichthyosaurs. After a century of paleontological investigation, what do we find?

> No earlier forms are known. The peculiarities of ichthyosaur structure would seemingly have required a long time for their development and hence a very early origin for the group, but there are no known Permian reptiles antecedent to them.[105]

Among the mammals there are a few aquatic groups. The pinnipeds (seals and sea lions) were thought to have evolved from one of the groups of carnivores during the late Eocene or Oligocene periods. But "because of the complete absence of any pre-Miocene fossils, it is not possible to indicate the actual ancestors of the pinnipeds."[106]

A last major group of mammals are the Cetacea (figure 13). Within the Cetacea are three major suborders (sometimes classified as orders) of aquatic mammals: the Archaeoceti (extinct toothed whales), the Odontoceti (porpoises, dolphins, beaked whales, and sperm whales), and the Mysticeti (whalebone whales). Each of these groups is highly specialized and shows little resemblance to any terrestrial mammals. In fact, because of this lack of resemblance (homology) and the absence of intermediates in the fossil record,[107] "the ancestry of the Cetacea is completely unknown."[108]

The Archaeoceti appear during the Eocene period. At a time when we would expect to find transitional forms, we find toothed whales that "were well specialized in many respects for pelagic life."[109] During the Miocene period, we find the Odontoceti. Are these transitional fossils? No, we find fossils, such as that of the porpoise in figure 13, that are "essentially modern in build" and are not led up to by a series of

Archaeocete
(BASILOSAURUS)

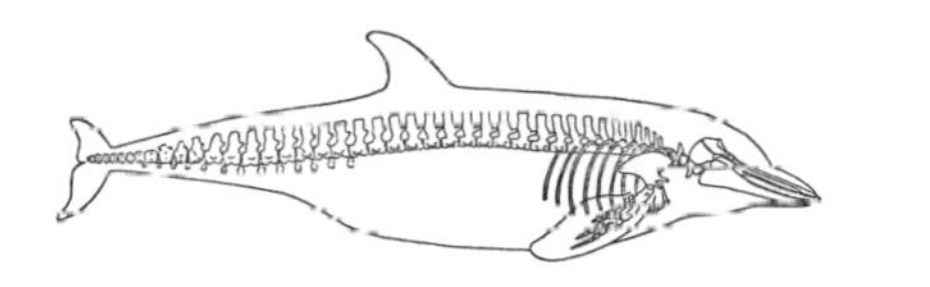

Odontocete
(KENTRIODON)

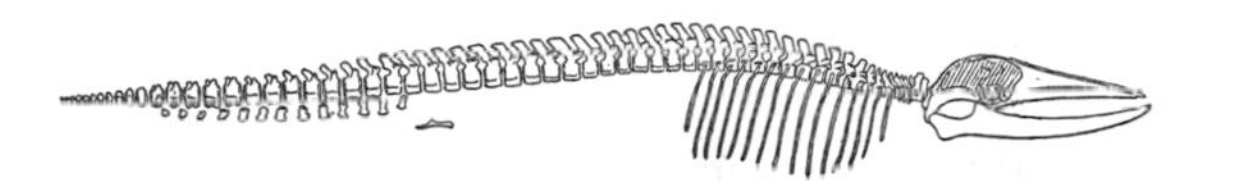

Mysticete
(BALAENOPTERA)

Figure 13. Bone structure of three groups of aquatic mammals.

transitional forms.[110] "Their previous geological record is as yet unknown."[111]

Mysticeti are said to have evolved from the Archaeoceti. However, the Mysticeti appear in the late Oligocene to late Miocene epochs, and their postulated ancestors lived in the Eocene. Between these groups there is an absence of any transitional forms.

An impartial observer would declare that Darwin's prediction concerning the future discovery of connecting links has not been fulfilled. The sharp discontinuities between the many varied animal types strongly contradicts the Neo-Darwinian model of evolution. In addition, the total absence of any transitional forms between major animal kinds raises many doubts about the Punctuated Equilibria model. The model that fits the facts most directly is the creation model.

Chapter Abstract

A cursory analysis is made of the fossil record of plants. The origin and evolutionary history of each group of modern plants including the flowering plants is not documented in the fossil record.

chapter seven

The Fossil Record of Plants

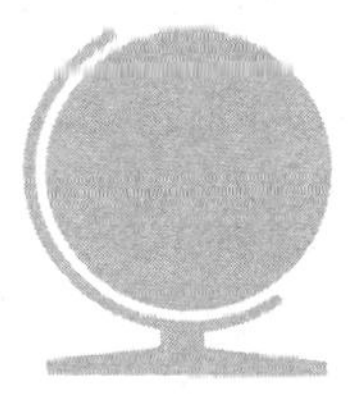

What has been said regarding the absence of ancestors and connecting links for the animal kingdom is even more pronounced for the plant kingdom. The basic groups or kinds of plants appear suddenly in the paleontological record and continue as distinct groups with major discontinuities between them.

It is not possible within the confines of this monograph to discuss in detail the postulated phylogenies (evolutionary trees) of the plants. Such schemes for the plants are even more speculative than they are for the animals.

The fossil record for all of the plants does not reveal a phylogenetic history that would be predicted by the models of evolution. As the late Chester Arnold of the University of Michigan stated in his book on paleobotany:

> It has long been hoped that extinct plants will ultimately reveal some of the stages through which existing groups have passed during the course of their development, but it

> must be freely admitted that this aspiration has been fulfilled to a very slight extent, even though paleontological research has been in progress for more than one hundred years. As yet we have not been able to trace the phylogenetic history of a single group of modern plants from its beginning to the present.[112]

The fossil record for plants is actually a much better case for creation than for evolution, as has been noted by E. J. H. Corner of the University of Cambridge:

> Much evidence can be adduced in favour of the theory of evolution — from biology, bio-geography, and paleontology, but I still think that, to the unprejudiced, the fossil record of plants is in favour of special creation.[113]

The Flowering Plants

Angiosperm (the flowering plants) fossils pose a particularly difficult problem because they appear suddenly in the Cretaceous layer and there is no trace of an ancestor in lower sediments. Since the time of Charles Darwin who called the situation "an abominable mystery," the problem has occupied the time and efforts of many paleobotanists. Percy Raymond described the situation this way in his book *Prehistoric Life:*

> These plants, dominant today, made a spectacular entry upon the world stage as great trees rather than as modest herbs, and paleobotanists are still ransacking the earth in search of their ancestors. The oldest known are singularly like forms now living. Their wood, leaves, flowers, and fruitification are of modern type. The two great groups, monocotyledons, with parallel venation (palms), and dicotyledons, with reticulate venation (common hardwoods), had already differentiated.[114]

In *An Introduction to Paleobotany,* Chester Arnold also describes this problem with great candor:

> Not only are plant evolutionists at a loss to explain the seemingly abrupt rise of the flowering plants to a place of dominance, but their origin is likewise a mystery.[115]

One of the most sensational discoveries in paleobotany has been that of microfossils, namely pollen and spores in lower Paleozoic rocks. Some paleobotanists still are unwilling to accept this information because of its staggering implications for evolution, but the evidences are too strong to ignore. Several scientists have reported the presence of gymnosperm (i.e., conifer) pollen as far down as the Cambrian period.[116] Previously, gymnosperms were not considered to have developed until the Devonian.

Plants called psilophytes (figure 14) were said to be ancestral to gymnosperms and other modern types of plants. However, the discovery of the pollen of seed-bearing plants in the Cambrian period has made it necessary to drastically revise evolutionary theory for

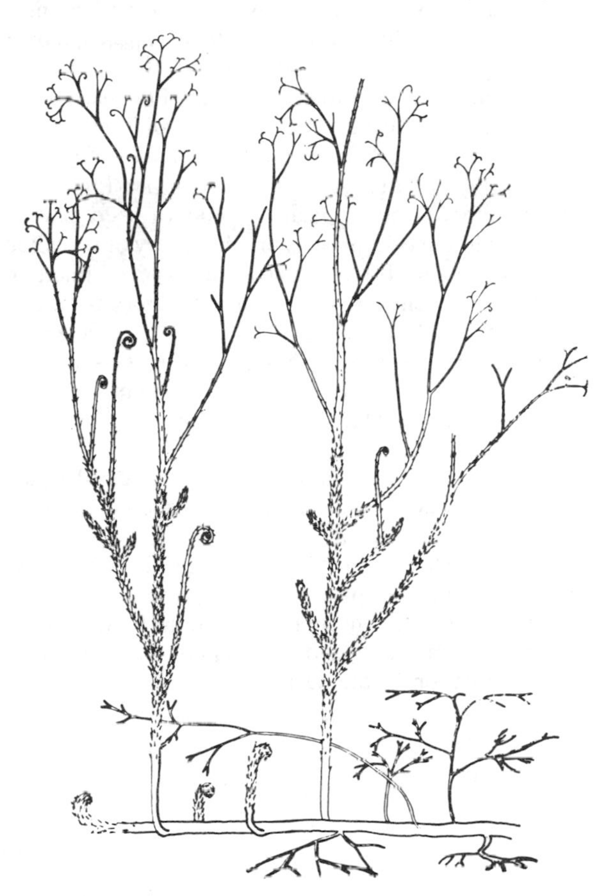

Figure 14. Structure of a fossil psilophyte.

plants. Daniel Axelrod has attempted to solve the problem by suggesting that the psilophytes should no longer be considered the ancestors of higher plants but a side branch in the plant phylogenetic tree. The appearance of complex microfossils of seed-bearing plants as low as the Cambrian means that the evolution of these forms must have occurred before the Cambrian period. The Precambrian, of course, has been unyielding to all our earnest efforts to find these plant ancestors. Thus the situation for plants is similar to that for animals, and the hope of finding ancestors is dim. As has been noted previously, many of these discontinuities tend to be accentuated by further collecting.

Frequently the literature of science describes some form of plant or animal farther down the geological column than was previously thought. For instance, a conifer cone from the lower Permian of West Texas contains plant embryos. These are said to be the oldest plant embryos on record.[117] What do they look like? These embryos are typical of modern seed plants. Almost without exception every discovery of so-called older specimens for any particular type of plant or animal only pushes that taxonomic category back into lower strata. It does not qualify as a connecting macroevolutionary step. In other words, the specimen is only another example of a known kind but in lower geological strata. The necessary transitional steps are still be be found.

A theory based too largely on the hopes of future discoveries is not well grounded. If we take the fossil record at face value, if we accept the sudden appearance of complex forms, if we accept the discontinuities between major groups as a valid representation of the past history of life, we are led toward the belief that the major types of plants and animals are the products of sudden creation in the past.

Chapter Notes

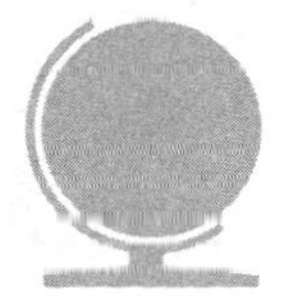

Chapter Abstract

Some concluding remarks are made concerning the persistent gaps in the fossil record. Each model of origins requires a faith commitment and the creation model is shown to be worthy of serious consideration.

chapter eight

Fitting the Pieces in the Paleopuzzle

Throughout this book, we have seen that the fossil record is characterized by a series of gaps between higher levels of classification. At the Precambrian-Cambrian border, there is a sudden outburst of highly complex life. According to the models of evolution, we should find simple ancestors leading up to these intricate and detailed organisms in the Cambrian period.

As we have examined the major groups of invertebrates, vertebrates, and plants, we have not been able to find transitional forms between higher levels of classification. "Nearly all categories above the level of families appear in the record suddenly and are not led up to by known, gradual, completely continuous transitional sequences."[118] And so we find that "gaps among known orders, classes, and phyla are systematic and almost always large."[119]

There is one more point that is worthy of our consideration. That is the question: When is a gap really a gap? In other words, are the gaps between the higher levels of classification in the fossil record real or imagined? How can we tell if there *were* or *were not* transitional forms in the past?

We can never positively answer that question. The probabilities that are calculated for burial, fossilization, survival, and recovery of any organism are only approximate. Thus, we cannot completely reconstruct the history of any group from its fossil record. We must be satisfied to attempt to reconstruct the original record as accurately as possible.

When Is a Gap Really a Gap?

Certainly there is evidence at present that would cause us to suspect that these gaps are real and not apparent. After more than a century of paleontological work we do not "need to apologize any longer for the poverty of the fossil record. In some ways it has become almost unmanageably rich, and discovery is out-pacing integration. . . ."[120]

Although we are finding more fossils, we are not significantly increasing the number of fossil species. This estimate was given by Norman D. Newell at the "Symposium on Fifty Years of Paleontology":

> Nearly fifty years ago, Schuchert (1910) estimated that there were one hundred thousand fossil species known. A recent tabulation by Muller and Campbell (1954) does not increase the earlier estimate. They find that ninety-two thousand species of fossil animals are known, and I suppose that the record of fossil plants would bring the total to around one hundred thousand.[121]

Although the total number of fossil species did not change during that fifty-year period, this is not to suggest that no new species were being discovered. In fact, this period was a very productive period in paleontology.

What was going on behind the scenes is most interesting. There was a silent revolution going on in classification. As new fossil finds were being discovered and named, older fossil finds were being reclassified into existing orders of classification. These fossils were not bridging the gaps between known orders of classification but instead were filling in these existing orders.

We can visualize this situation by considering a two-dimensional graph which relates the number of fossil finds to the anatomical structure of each of these fossils (figure 15). In this graph, we have chosen to express structure in discrete, quantitative intervals (i.e., increase in bone length, increase in shell curvature). Each of the different models for origins would be represented by a different graphical distribution.

The Neo-Darwinian model of evolution would predict that chance mutations arising in a population of organisms would eventually achieve dominance in a population. It would take numerous mutations to develop even one *new* species. This process would be so lengthy that there would be an abundance of transitional forms. Although we would expect to find more of some fossils than others (because they were more abundant or were more easily fossilized), we would certainly expect to find a number of transitional forms that bridge the gaps between each fossil find.

The Punctuated Equilibria model would not predict that we would find an abundance of transitional forms. However, even organisms that are evolving in isolated populations would have some probability of fossilization. Occasionally, some fossils would be fossilized and discovered to provide some "stepping stones" in the valleys between the hills of fossil finds.

This prediction is even more reasonable when we realize that it would take numerous species transitions to bridge the gaps between organisms that have evolved major adaptive structures (wings, flippers). According to Steven Stanley of Johns Hopkins University, "ten or so species-to-species phyletic transitions are insufficient to produce the enormous degree of change that occurred in the origin of such divergent taxa as bats and whales."[122]

These species-to-species transitions would consist of numerous populations over time. Each of these populations would be composed of hundreds, thousands, or perhaps millions of individuals. It is certainly curious that we have not found occasional fossil intermediates.

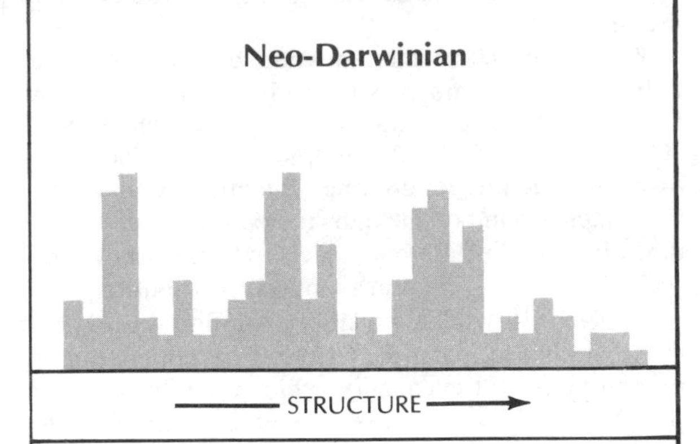

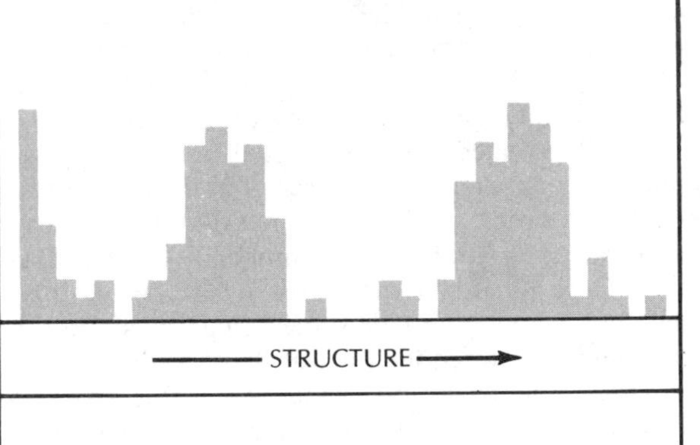

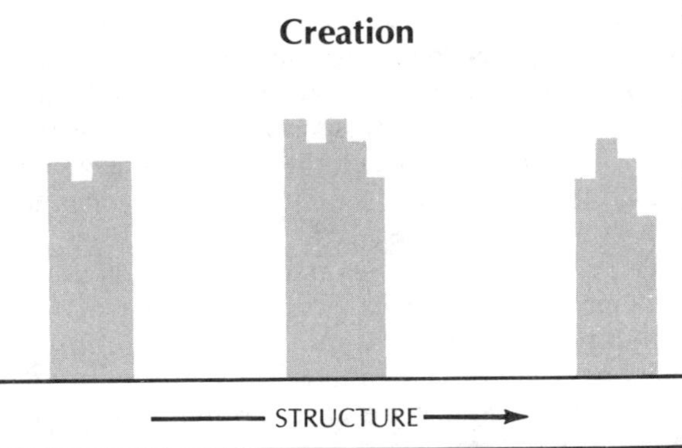

Figure 15. The predictions made by three models concerning gaps in the fossil record.

As we have considered the fossil record, we have noted occasional fossil finds which could be considered intermediates. However, upon further examination, we have found that those which were proposed as transitional forms (*Peripatus, Seymouria,* nothosaurs) were not. Similarity in structure in these cases did not indicate common ancestry. The intermediate links that we should be able to find, we have not.

As we continue to collect fossils from the geologic column, we continue to find species that fill out our existing levels of classification rather than bridge the gaps between them. "Experience shows that the gaps which separate the highest categories may never be bridged in the fossil record. Many of the discontinuities tend to be more and more emphasized with increased collecting."[123]

What we find in the fossil record are these gaps which look like canyons and fossil numbers which appear as cliffs on either side of these canyons. These discontinuities are being more emphasized because we are building the cliffs higher while the canyons remain empty. It might be the hope of evolutionists that future discovery of fossil finds will "plug these gaps." But, "most paleontologists and stratigraphers would predict that no amount of future field work will ever fill a majority of existing phyletic gaps between transient species."[124] It would seem from this information as well as from our investigation of the fossil record that the creation model is certainly worthy of serious consideration.

As has been noted previously in this book, it is not possible to prove any of the models of origins scientifically. They are outside of empirical scientific methods for testing their validity. In order to accept any of these models, we must exercise some degree of faith.

In the face of this evidence, why is it that a majority of the scientific community has been unwilling to consider the creation model? The answer may lie in the presuppositions of the observers rather than in the facts that are available. Many paleontologists feel that it is scientifically required that they continue to work within the confines of the evolutionary model rather than to consider alternative models.

> Almost all paleontologists recognize that the discovery of a complete transition is in any case unlikely. Most of them find it logical, if not scientifically required, to assume that the sudden appearance of a new systematic group is not evidence for special creation or for saltation, but simply means that a full transitional sequence more or less like those that are known did occur and simply has not been found in this instance.[125]

In some cases, evolutionists have been unwilling to consider the creation model because they feel that it requires too great of a faith commitment. As paleontologist D. M. S. Watson has stated, the theory of evolution is accepted "not because it can be proved by logically coherent evidence to be true but because the only alternative, special creation, is clearly incredible."[126]

Could we also ask, Are the models of evolution incredible? They require us to believe that nonliving substances gave rise to living creatures some time in the past through some undiscovered process and against spectacular mathematical odds.[127] It requires that we believe these initial forms of life evolved into more complex forms of life. Neither the initial generation of life nor the subsequent proliferation of major types can be demonstrated by laboratory experimentation or paleontological investigation.

The creation model requires that we believe life is a result, not of naturalistic processes, but of direct intervention from a supernatural Creator. In other words, it requires that we believe that life gave rise to life and that a Creator gave rise to His creation. It is a faith commitment that is based upon objective evidence from the fossil records.

As we continue our investigation of the fossil records, it would seem only reasonable scientifically and fair academically if we were to consider various models of origins in our analysis of the data. We believe that the creation model should be given serious consideration.

But the creation model can be supported by more than evidence from the field of science. It is part of a broad, unifying, and coherent world view. It can be substantiated by the authority of the Bible, which claims to be the direct revelation of God. Also we can check the direct testimony of a person who claimed to be God and who made reference to the act of creation as an event in history.

If we can check the authority of these two witnesses and if we find them to be reliable, then we have substantial evidence for the creation model. These statements and claims can be checked out through standard methods of investigation that are used in the fields of history, philosophy, religion, archaeology, and science. The application of these tests to this written revelation (the Bible) and physical revelation (Jesus of Nazareth) will indicate to the inquirer that there is indeed a Creator and that He has spoken to man.

The most up-to-date evidence that we have from various disciplines combines with biblical information to declare the validity of the first five words of the Bible:

> In the beginning God created . . .

Response

RUSSELL L. MIXTER

Widespread knowledge of the literature on the occurrence of fossils has led these authors to collect much information to bolster their belief that the gaps in the paleontological record are real and therefore evidence that there must have been a creation of many kinds of organisms, rather than derivation of them from previously existing forms. Evolutionists say the links are missing because they have never been found, or would not have been fossilized anyhow, or have been destroyed by geological processes, or are resting on someone's shelf and have not yet been described. This monograph maintains that because the Bible teaches specific creations, there would therefore be no continuity of fossils from one kind to another. Here is a well-documented discussion of the creationist's position.

When Darwin visited the Galapagos Islands off the west coast of South America, he observed fourteen species of finches, birds with bills like sparrows. However, some finches had bills that were used to hold cactus spines for piercing insects in bark, others had more fragile bills similar to those of warblers, and still others had heavy bills for devouring cactus or seeds. Because the bird most similar to these island finches was a South American resident, Darwin reasoned that the continental bird had migrated to the islands and diversified into several genera and species as it spread over the areas. The reader will note that this present volume agrees to the reasonableness of such a conclusion by saying "change is possible within limits. In fact, as we look at the fossil record, the results from genetic research, and the natural world about us, we are led to believe that the truth lies between the two extremes of fixity of species and limitless change."

The idea of fixity of species is not from Scripture but started with Linnaeus, who saw a limited group of distinct species in Scandinavia and concluded that whatever forms he identified were obviously all created species. Later when he saw more of the world's specimens, he reasoned that the several species within a genus could have come from the originally created species. But some Christians have adopted the first view that Linnaeus held and maintain even today that no species will evolve into another.

The biblical phrase "after its kind" could be paraphrased as "kinds of." Hence, where the Bible says, "Let the earth bring forth grass . . . after its kind," it implies that several types appeared, but does not state that the first ones created must remain just as they were when they first appeared. So one may study any group of plants or animals living today or represented by fossil remains and draw whatever conclusion he considers reasonable as to the amount of change that has occurred since creation.

Because the gaps in the record of the rocks occur between orders of organisms, it seems clear that no order of animals or plants has been derived from any other. To illustrate, the marsupials, represented by the kangaroo and opossum, are not connected by a graded series of fossils to any other order, such as the edentates to which armadillo and anteater belong. Logically then, one may say that within an order, the varied species may have come from a common ancestor, but that the missing links testify to special creations of the first members of the orders.

This book is commendable for accepting the conclusions of paleontologists concerning which fossils are found in certain strata. If one feels this treatise is using the argument from silence, recall that Scripture implies that there would have been silences, that is, gaps in the record as missing links. Read this book, then, to get a clearly presented view by two well-read and judicious writers.

References

[1]Dwight Davis, "Comparative Anatomy and the Evolution of the Vertebrates," in *Genetics, Paleontology, and Evolution,* ed. G. L. Jepsen, G. G. Simpson, E. Mayr (Princeton, N.J.: Princeton University Press, 1949), pp. 64–89; George Howe, "Homology, Analogy, and Creative Components in Plants," in *Scientific Studies in Special Creation,* ed. Walter Lammerts (Nutley, N.J.: Presbyterian & Reformed, 1971).

[2]Richard E. Dickerson, "The structure of cytochrome c and the rates of molecular evolution," in *Molecular Evolution* (1971): 26–45; James F. Coppedge, *Evolution: Possible or Impossible?* (Grand Rapids: Zondervan, 1973), pp. 203–204.

[3]Theodosius Dobzhanzky, *Genetics and the Origin of Species,* 3rd ed. (New York: Columbia University Press, 1951); Bolton Davidheiser, *Evolution and Christian Faith* (Nutley, N.J.: Presbyterian & Reformed, 1969), pp. 187–216.

[4]Ernst Mayr, *Populations, Species, and Evolution* (Cambridge: Harvard University Press, Belknap Press, 1970); Walter E. Lammerts, "The Galapagos Island Finches," in *Why Not Creation?* ed. Walter Lammerts (Nutley, N.J.: Presbyterian & Reformed, 1970), pp. 354–366.

[5]P. S. Moorhead and M. M. Kaplan, eds., *Mathematical Challenges to the Neo-Darwinian Interpretation of Evolution* (Philadelphia: Wistar Institute, 1967); Harold F. Blum, *Time's Arrow and Evolution* (Princeton, N.J.: Princeton University Press, 1968). D. E. Hull, "Thermodynamics and Genetics of Spontaneous Generation," *Nature* 186 (1960): 693-694.

[6]Coppedge, *Evolution: Possible or Impossible?* pp. 19–182; Duane T. Gish, *Speculations and Experiments Related to Theories on the Origin of Life* (San Diego: Creation-Life Publishers, 1972); Bradley, Olsen, and Thaxton, *Life: The Crisis in Chemistry* (future Christian Free University book).

[7]Carl O. Dunbar, *Historical Geology* (New York: Wiley, 1949), p. 52.

[8]George Gaylord Simpson, *The Meaning of Evolution* (New York: Bantam Books, 1967), p. 124.

[9]David B. Kitts, "Paleontology and Evolutionary Theory," *Evolution* 28 (1974): 466.

[10]George Gaylord Simpson, "The Present Status of the Theory of Evolution," *Proceedings of the Royal Society of Victoria* 82, no. 2 (1969): 149–160.

[11]Simpson, *Meaning of Evolution,* p. 211.

[12]George Gaylord Simpson, "The History of Life," in *Evolution After Darwin,* vol. 1 of *The Evolution of Life,* ed. Sol Tax (Chicago: University of Chicago Press, 1960), p. 149.

[13]Kitts, "Paleontology and Evolutionary Theory," p. 466.

[14]Charles Darwin, *The Origin of Species by Means of Natural Selection* (1859; reprint ed., New York: Collier Books, 1962), p. 308.

[15]Norman D. Newell, "The Nature of the Fossil Record," *Proceedings of the American Phil. Society* 103, no. 2 (1959): 264–285.

[16]Richard B. Goldschmidt, *The Material Basis of Evolution* (New Haven: Yale University Press, 1940).

[17]O. H. Schindewolf, *Grundfragen der Paleontologie* (Stuttgart: E. Schweizerbarthsche Verlagsbuchhandlung, 1950).

[18]Richard B. Goldschmidt, "Evolution, as Viewed by One Geneticist," *American Scientist* 40 (1952): 84–98,135.

[19]George Gaylord Simpson, *Tempo and Mode in Evolution* (New York: Columbia University Press, 1944); Kitts, "Paleontology and Evolutionary Theory," p. 466; Garrett Hardin, *Nature and Man's Fate* (New York: Holt, Rinehart & Winston, 1959), pp. 261–262.

[20]Ernst Mayr, *Animal, Species, and Evolution* (Cambridge: Harvard University Press, Belknap Press, 1963).

[21]Niles Eldredge and Stephen J. Gould, "Punctuated Equilibria: An Alternative to Phyletic Gradualism," in *Models in Paleobiology,* ed. T. J. M. Schopf (San Francisco: Freeman Cooper, 1972).

[22]John C. Whitcomb and Henry M. Morris, *The Genesis Flood* (Grand Rapids: Baker, 1961); Duane T. Gish, *Evolution? The Fossils Say No!* (San Diego: Creation-Life Publishers, 1974); Bernard Ramm, *The Christian View of Science and Scripture* (Grand Rapids: Eerdmans, 1954); Russell L. Mixter, *Creation and Evolution* (Mankato: The American Scientific Affiliation, 1963).

[23]Ernst Mayr, *Animal, Species, and Evolution* (Cambridge: Harvard University Press, Belknap Press, 1963).

[24]John Imbrie, "The Species Problem With Fossil Animals," in *The Species Problem*, ed. Ernst Mayr (Washington, D.C.: American Association for the Advancement of Science, 1957).

[25]W. B. Harland et al., eds. *The Fossil Record* (London: Geological Society, 1967).

[26]Norbert E. Cygan and Frank L. Koucky, "The Cambrian and Ordovician Rocks of the East Flank of the Big Horn Mountains, Wyoming," in *Wyoming Geological Association and Billings Geological Society Guidebook* (Northern Powder River Basin, 1963).

[27]Darwin, *Origin of Species*, pp. 331–332.

[28]Newell, "Nature of the Fossil Record," pp. 264–285.

[29]J. Marvin Weller, *The Course of Evolution* (New York: McGraw-Hill, 1969).

[30]Heinz Dombrowski, "Bacteria From Paleozoic Salt Deposits," *Acad. Sci. Ann.* 108 (New York, 1963): 453–460. Dorothy Zeller Oehler, "Transmission Electron Microscopy of Organic Microfossils From the Late Precambrian Bitter Springs Formation of Australia: Techniques and Survey of Preserved Ultrastructure," *Journal of Paleontology* 50 (1976): 90–106; Barbara Javor and Eric Mountjoy, "Late Proterozoic Microbiota of the Miette Group, Southern British Columbia," *Geology* 4 no. 2 (1976): 111–119.

[31]Martin F. Glasessner, "Pre-Cambrian Animals," *Scientific American* 204, no. 3 (1961): 72–78.

[32]Preston Cloud, "Pseudofossils: A Plea for Caution," *Geology* 1, no. 3 (1973): 123–127.

[33]Daniel I. Axelrod, "Early Cambrian Marinefauna," *Science* 128 (1958): 7–9.

[34]A. H. Clark, *The New Evolution: Zoogenesis* (Baltimore: Williams and Wilkins, 1930); Richard Goldschmidt, *The Material Basis of Evolution* (New Haven: Yale University Press, 1940).

[35]M. T. Ghiselin, "Models of Phylogeny," in *Models in Paleobiology*, ed. T. J. M. Schopf (San Francisco: Freeman Cooper, 1972).

[36]G. A. Kerkut, *Implications of Evolution* (New York: Pergamon Press, 1960), p. 47.

[37]J. R. Baker, "The Status of the Protozoa," *Nature* 161 (1948): 548–587; A. C. Hardy, "On the Origin of the Metazoa," *Quart. Jour. Micr. Sci.* 84 (1953): 441.

[38]E. Ray Lankester, "Introduction and Protozoa. Fascicle 1," in *A Treatise on Zoology* (London: Adam and Charles Black, 1909).

[39]Kerkut, *Implications of Evolution,* p. 49.

[40]Ibid., pp. 50–100.

[41]Raymond Moore, Cecil Lalicker, and Alfred Fischer, *Invertebrate Fossils* (New York: McGraw-Hill, 1952).

[42]Newell, "Nature of the Fossil Record," pp. 264–285.

[43]Percy E. Raymond, *Prehistoric Life* (Cambridge: Harvard University Press, 1958).

[44]Carl O. Dunbar, *Historical Geology* (New York: Wiley, 1964).

[45]James R. Beerbower, *Search for the Past* (Englewood Cliffs, N.J.: Prentice-Hall, 1960).

[46]Clark, *New Evolution: Zoogenesis,* p. 112.

[47]G. E. Hutchinson, "Restudy of Some Burgess Shale Fossils," *U. S. Natural Museum Proceedings* 78, art. 11 (1930); Charles D. Walcott, "Addenda to Descriptions of Burgess Shale Fossils," Smithsonian Misc. Collection 85, no. 3 (1931): 1–46.

[48]B. F. Howell, "Worms," in *Treatise on Invertebrate Paleontology,* Part W, ed. Raymond C. Moore et al. (Geol. Soc. Am. and U. Kansas Press, n.d.).

[49]Clark, *New Evolution: Zoogenesis,* p. 100.

[50]Kerkut, *Implications of Evolution,* pp. 112–129.

[51]J. Hadzi, *The Evolution of the Metazoa* (New York: Macmillan, 1963).

[52]Raymond, *Prehistoric Life,* p. 87.

[53]Cygan and Koucky, "Cambrian and Ordovician Rocks."

[54]F. D. Ommaney, *The Fishes* (New York: Time-Life, Life Nature Library, 1964), p. 60.

[55]N. J. Berril, *The Origin of Vertebrates* (Oxford University Press, 1955), p. 10.

[56]Alfred S. Romer, *Vertebrate Paleontology* (Chicago: University of Chicago Press, 1966), pp. 15–16.

[57]Ibid., pp. 15,24.

[58]Ibid., p. 33.

[59]Ibid., pp. 34,38.

[60]Kerkut, *Implications of Evolution,* pp. 52–53.

[61]Romer, *Vertebrate Paleontology,* pp. 52–53.

[62]Erik Jarvik, "On the Structure of the Snout of Crossopterygians and Lower Gnathostomes in General," *Zool. Bidr.* 21 (1942): 235.

[63]Keith S. Thompson, "The Ancestry of the Tetrapods," *Science Progress* 52 (1964): 451–459.

[64]Erik Jarvik, "The Oldest Tetrapods and Their Forerunners," *Science Monthly* 80 (1955): 141–154; G. Save-Soderbergh, "Preliminary Note on Devonian Stegocephalians From East Greenland," *Medd. Grønland,* vol. 94, no. 7 (1932).

[65]Theodore H. Eaton, "The Aquatic Origin of the Tetrapods," *Trans. Kansas Acad. Sci.* 63 (1960): 115–120.

[66]T. S. Westoll, "The Origin of the Tetrapods," *Biol. Rev. Cambridge Phil. Soc.* 18 (1943): 78–98.

[67]T. S. Westoll, "The Origin of the Primitive Tetrapod Limb," *Proc. Royal Society of London,* B 131 (1943): 373–393.

[68]Bobb Schaeffer, "The Rhipidistian-Amphibian Transition," *Am. Zool.* 5 (1965): 267–276.

[69]Alfred S. Romer, "Tetrapod Limbs and Early Tetrapod Life," *Evolution* 12 (1958): 365–369.

[70]Gordon Gunter, "Origin of the Tetrapod Limb," *Science* 123 (1956): 495–496.

[71]G. L. Orton, "Original Adaptive Significance of the Tetrapod Limb," *Science* 120 (1954): 1042–1043.

[72]D. W. Ewer, "Tetrapod Limb," *Science* 122 (1955): 467–468; C. J. and O. B. Goin, "Further Comments on the Origin of the Tetrapods," *Evolution* 10 (1956): 440–441.

[73]F. E. Warburton and N. S. Denman, "Larval Competition and the Origin of Tetrapods," *Evolution* 15 (1961): 566.

[74]R. F. Inger, "Ecological Aspects of the Origin of the Tetrapods," *Evolution* 11 (1957): 373–376.

[75]Romer, *Vertebrate Paleontology,* p. 98.

[76]M. K. Hecht, "A Reevaluation of the Early History of the Frogs," *Syst. Zool.* 11, no. 1 (1962): 39–44.

[77]T. S. Parsons and E. E. Williams, "The Relationships of the Modern Amphibia: A Reexamination," *Quart. Rev. Biol.* 38 (1963): 26–53.

[78]Edwin H. Colbert, *Evolution of the Vertebrates* (New York: Wiley, 1969), p. 111.

[79]Kerkut, *Implications of Evolution,* p. 136.

[80]G. Heilman, *The Origin of Birds* (London: Witherby, 1926); G. R. deBeer, *Archaeopteryx Lithographica* (London: British Museum of Natural History, 1954); Adrian J. Desmond, *The Hot-Blooded Dinosaurs: A Revolution in Paleontology* (New York: Dial, 1976).

[81]Harvey I. Fisher, "The Occurrence of Vestigial Claws on the Wings of Birds," *Am. Mid. Nat.* 23 (1940): 234–243.

[82]N. Kitchen Parker, "On the Structure and Development of the Wings in the Common Fowl," *Phil. Trans. Roy. Soc.,* B, vol. 179 (London, 1889).

[83]N. Kitchen Parker, "On the Morphology of a Reptilian Bird *Opisthocomus cristatus,*" *Trans. Zool. Soc., London,* vol. 13, pt. 2, no. 1 (1891).

[84]Frank W. Cousins, "The Alleged Evolution of Birds (*Archaeopteryx*)," in *A Symposium on Creation,* vol. 3, ed. Donald Patten (Grand Rapids: Baker, 1971).

[85]Heilman, *Origin of Birds.*

[86]John H. Ostrom, "The Origin of Birds," *Ann. Rev. Earth Plan. Sci.* 3 (1975): 55–77.

[87]William A. Gregory, "Theories on the Origin of Birds," *Acad. Sci. Ann.* 27 (New York, 1916): 31–38.

[88]Ostrom, "Origin of Birds."

[89]W. E. Swinton, "The Origin of Birds," in *Biology and Comparative Physiology of Birds,* ed. Marshall (London: Academic, 1960), p. 1.

[90]Leigh VanValen, "Therapsids As Mammals," *Evolution* 14 (1960): 303–313.

[91]Simpson, *Tempo and Mode in Evolution,* p. 361.

[92]E. C. Olson, *The Evolution of Life* (London: Weidenfeld and Nicolson, 1965), p. 175.

[93]F. M. Carpenter, "The Geological History and Evolution of Insects," *American Scientist* 41 (1953): 256–270.

[94]Olson, *Evolution of Life,* p. 175.

[95]Ibid., p. 176.

[96]John H. Ostrom, "*Archaeopteryx* and the Origin of Flight," *Quart. Rev. Biol.* 49 (1974): 27–47.

[97]Olson, *Evolution of Life,* p. 176.

[98]Glenn L. Jepsen, "Early Eocene Bat From Wyoming," *Science* 154 (1966): 1333–1339.

[99]Glenn L. Jepsen, "Bat Origins and Evolution," in *Biology of Bats*, vol. 1, ed. William A. Wimsatt (New York: Academic, 1970), p. 12.

[100]Olson, *Evolution of Life*, p. 173.

[101]Romer, *Vertebrate Paleontology*, p. 116.

[102]Ibid., p. 117.

[103]Ibid., p. 126.

[104]Ibid., p. 125.

[105]Ibid., p. 120.

[106]Colbert, *Evolution of the Vertebrates*, p. 361.

[107]S. E. Kleinenberg, "On the Origin of Cetacea," *Proc. 15th Int. Cong. Zool.* (1958), pp. 445–447.

[108]Colbert, *Evolution of the Vertebrates*, p. 340.

[109]Remington Kellogg, "The History of Whales — Their Adaptation to Life in the Water," *Quart. Rev. Biol.* 3 (1928): 29–76.

[110]Romer, *Vertebrate Paleontology*, p. 116.

[111]Kellogg, "History of Whales," p. 46.

[112]Chester A. Arnold, *An Introduction to Paleobotany* (New York: McGraw-Hill, 1947), p. 7.

[113]E. J. H. Corner, "Evolution," in *Contemporary Botanical Thought*, ed. Anna M. MacLeod and L. S. Copley (Chicago: Quadrangle Books, 1961).

[114]Raymond, *Prehistoric Life*, p. 298.

[115]Arnold, *Introduction to Paleobotany*, p. 334.

[116]S. Leclercq, "Evidence of Vascular Plants in the Cambrian," *Evolution* 10 (1956): 109–114; Daniel I. Axelrod, "Evolution of the Psilophyte Paleoflora," *Evolution* 13 (1959): 264–275.

[117]C. N. Miller and J. T. Brown, "Paleozoic Seeds With Embryos," *Science* 179 (1973): 184–185.

[118]George Gaylord Simpson, *The Major Features of Evolution* (New York: Columbia University Press, 1953), p. 360.

[119]Simpson, "History of Life."

[120]T. Neville George, "Fossils in Evolutionary Perspective," *Science Progress* 48 (1960): 1–30.

[121]Newell, "Nature of the Fossil Record," pp. 264–285.

[122]Steven M. Stanley, "Stability of Species in Geologic Time," *Science* 192 (1976): 267–268.

[123] Norman D. Newell, "Adequacy of the Fossil Record," *Journal of Paleontology* 33 (1959): 488–499.

[124] Imbrie, "Species Problem With Fossil Animals."

[125] Simpson, *Major Features of Evolution,* p. 360.

[126] D. M. S. Watson, "Adaptation," *Nature* 124 (1929): 233.

[127] Bradley, Olsen, and Thaxton, *Life: The Crisis in Chemistry;* Coppedge, *Evolution: Possible or Impossible?*

Glossary

Allopatric	*Populations of organisms originating in or occupying two or more different geographic areas*
Angiosperms	*Plants of the class Angiospermae, which comprises flowering plants with seeds that are enclosed by an ovary*
Annelids	*Worms belonging to the phylum Annelida having a body made of joined segments*
Arthropods	*Invertebrate animals of the phylum Arthropoda with jointed legs and segmented bodies (insects, trilobites, crustaceans)*
Brachiopods (brack-key-o-pods)	*Invertebrate marine animals belonging to the phylum Brachiopoda; having hinged upper and lower shells enclosing two armlike parts with tentacles*
Calcareous (kal-ker-e-us)	*Containing calcium carbonate, calcium, or lime*
Cambrian	*The first geologic period in the Paleozoic Era, characterized by an abundance of marine multicellular animals*
Chitinous (kite-tin-us)	*Containing chitin, a tough compound secreted by the outer layer of skin of many invertebrates*
Coelenterates (si-len-te-rates)	*Invertebrate animals belonging to the phylum Coelenterata, such as jellyfishes and corals*
Competition	*The interaction that exists between two or more organisms when the requirements of one or more of the organisms living in a community cannot be obtained from the available supply of resources*
Cretaceous	*The third period of the Mesozoic Era, characterized by an abundance of reptiles*
Fauna	*The aggregate of all the kinds of animals in an area or in a geological period*

Genotype	*The fundamental hereditary constitution (the sum total of genes) of an organism*
Hemichordates	*Marine invertebrates of the phylum Hemichordata with visceral clefts and with suggestions of a simple notochord and a dorsal nervous system*
Metazoa	*All multicellular animals with differentiated cells and organs*
Mollusks	*Invertebrate animals belonging to the phylum Mollusca, such as clams, snails, and squid*
Niche	*(1) ecological niche: the role of an organism in the environment, its activities and relationships to the biotic and abiotic environment* *(2) habitat niche: the smallest unit of a habitat occupied by an organism*
Ontogeny	*The biological development of the individual*
Ordovician (or-do-*vish*-on)	*The second period of the Paleozoic Era immediately following the Cambrian period*
Ossification	*The development of bony material*
Phenotype	*The manifest characteristics of an organism collectively, including the anatomical and physiological traits*
Phylogeny	*The evolutionary lines of descent of any plant or animal species*
Planula (*plan*-yoo-la)	*A ciliated free-swimming larvae*
Precambrian	*The earliest geological time, covering all time previous to the Cambrian period*
Predator	*An animal that attacks and eats another animal*
Psilophytes (*sigh*-lo-fights)	*Extinct plants found in the Devonian period that possess true vascular (phloem and xylem) tissue*
Siliceous (sil-*lish*-us)	*Containing silica, a hard glassy mineral*
Syncitial (sin-sish-al)	*A mass of cytoplasm, enclosed in a single continuous plasma membrane, containing many nuclei*

Geologic Nomenclature
(listed in sequential order)

I. Cenozoic Era

A. Quaternary Period
1. Recent
2. Pleistocene Epoch

B. Tertiary Period
1. Pliocene Epoch
2. Miocene Epoch
3. Oligocene Epoch
4. Eocene Epoch
5. Paleocene Epoch

II. Mesozoic Era

A. Cretaceous Period
B. Jurassic Period
C. Triassic Period

III. Paleozoic Era

A. Permian Period
B. Carboniferous Period
1. Pennsylvanian Period
2. Mississippian Period
C. Devonian Period
D. Silurian Period
E. Ordovician Period
F. Cambrian Period

IV. Precambrian Time